Gain-Cell Embedded DRAMs for Low-Power VLSI Systems-on-Chip

Pascal Meinerzhagen • Adam Teman
Robert Giterman • Noa Edri • Andreas Burg
Alexander Fish

Gain-Cell Embedded DRAMs for Low-Power VLSI Systems-on-Chip

Springer

Pascal Meinerzhagen
Intel Labs, Circuit Research Lab
Intel Corporation
Hillsboro, Oregon, USA

Adam Teman
Faculty of Engineering
Bar-Ilan University
Ramat Gan, Israel

Robert Giterman
Faculty of Engineering
Bar-Ilan University
Ramat Gan, Israel

Noa Edri
Faculty of Engineering
Bar-Ilan University
Ramat Gan, Israel

Andreas Burg
EPFL STI IEL TCL
Lausanne, Switzerland

Alexander Fish
Faculty of Engineering
Bar-Ilan University
Ramat Gan, Israel

ISBN 978-3-319-60401-5 ISBN 978-3-319-60402-2 (eBook)
DOI 10.1007/978-3-319-60402-2

Library of Congress Control Number: 2017943168

Printed on acid-free paper

This Springer imprint is published by Springer Nature
The registered company is Springer International Publishing AG
The registered company address is: Gewerbestrasse 11, 6330 Cham, Switzerland

To Runa and Elias

To Hadas and Shalev

To Tom, Daniel and Jonathan

Contents

Chapter 1
Embedded Memories: Introduction

1.1 Increasing Need for Embedded Memories in Low-Power VLSI SoCs

There is a steadily increasing need for *embedded memories* in very large scale integration (VLSI) system-on-chip (SoC) designs targeted at microprocessors (used in servers, personal computers, laptop computers, tablets, and smartphones), biomedical implants, wireless communications systems, and many other applications. Such embedded memories are required to temporarily store data and/or instructions. From a system level perspective, it is clearly advantageous to always have more memory embedded directly on the compute chip, rather than relying on external memory chips. The primary reasons for this are: (1) embedded memories allow higher system-level integration densities, and (2) going off-chip through input/output (I/O) pads and capacitive lines on printed circuit boards (PCBs) entails severe speed and power penalties compared to on-chip connections [12]. As shown in Fig. 1.1a, the total cache size requirement in microprocessors has increased by around 5× in a time interval as short as 4 years. In fact, back in 2005, an Intel® Pentium® D microprocessor used around 2 MB of cache memory, while the Intel® Core™i7, released in 2009, takes advantage of almost 10 MB of cache memory [17]. In accordance with this past trend of quickly increasing demand for embedded memories, the International Technology Roadmap for Semiconductors (ITRS) predicted in its 2011 Edition that the total embedded memory size for general SoC applications will increase by almost 50× over the next 15 years [10], as shown in Fig. 1.1b.

As of today, embedded memories typically consume at least 50% of the total area and power budget of VLSI SoCs [10]. Figure 1.2 illustrates this by showing the layout pictures and the chip microphotographs of various VLSI SoCs, ranging from high-end microprocessors, through wireless communications systems, to ultra-low power (ULP) subthreshold (sub-V_T) microprocessors for health monitoring. The embedded memories, in the form of static random-access memory (SRAM)

P. Meinerzhagen et al., *Gain-Cell Embedded DRAMs for Low-Power VLSI Systems-on-Chip*, DOI 10.1007/978-3-319-60402-2_1

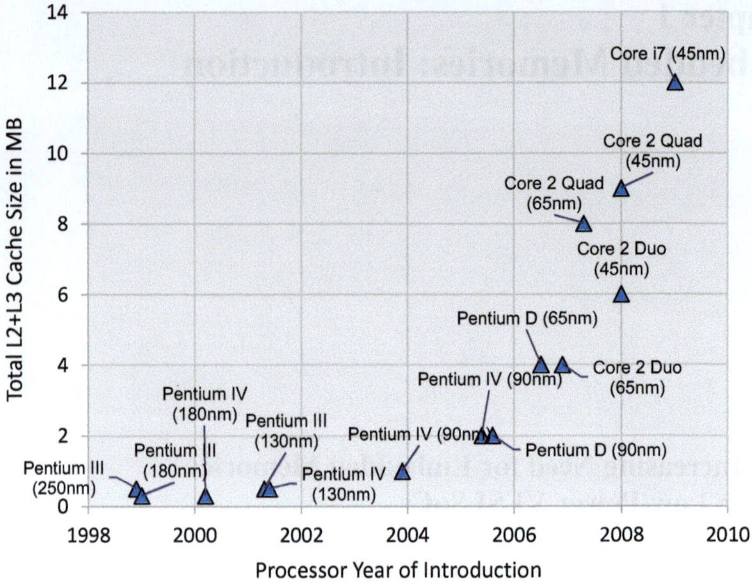

(a) Evolution of total cache size in microprocessors since 1998 [1].

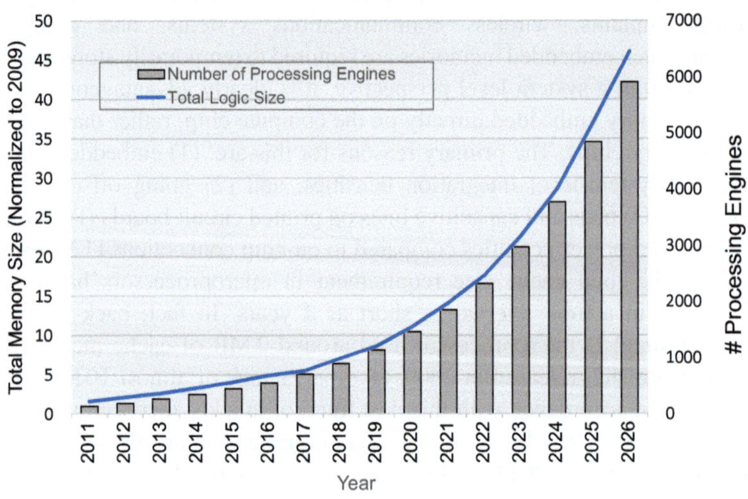

(b) Predicted evolution of total memory size in SoCs [2].

Fig. 1.1 (**a**) Past evolution, and (**b**) predicted future evolution of embedded memory size

macrocells, are visible as regular layout tiles. Especially in case of the sub-V_T microprocessor, shown in Fig. 1.2d, the embedded memories, visible as yellow tiles, consume a dominant area share compared to the logic core which is in the center

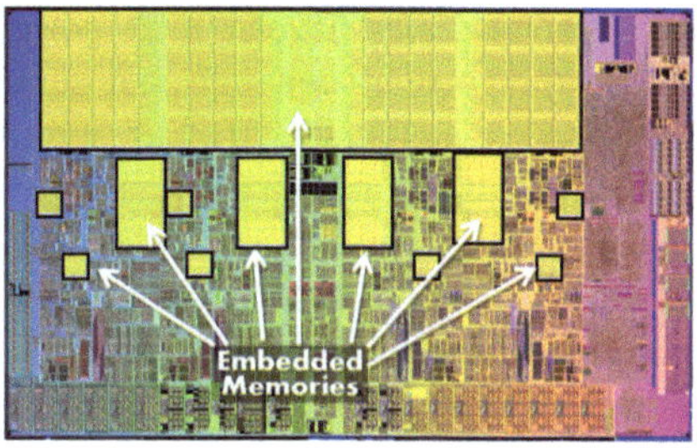

(a) Layout picture of 45 nm Intel® Core™ i7 processor (Nehalem) [4].

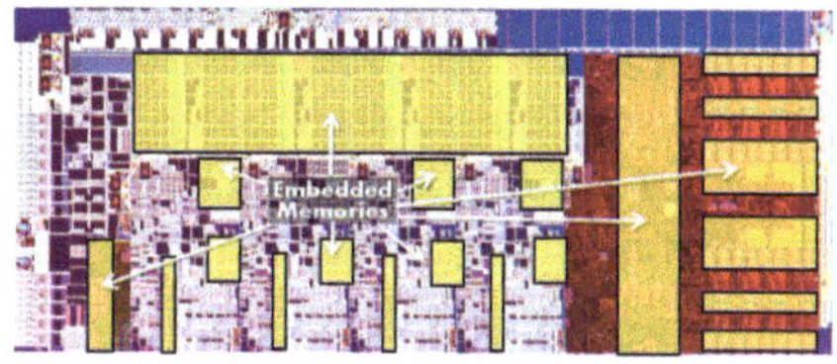

(b) Layout picture of 22 nm Intel® processor (a multi-CPU and GPU SoC) codenamed Ivy Bridge [9].

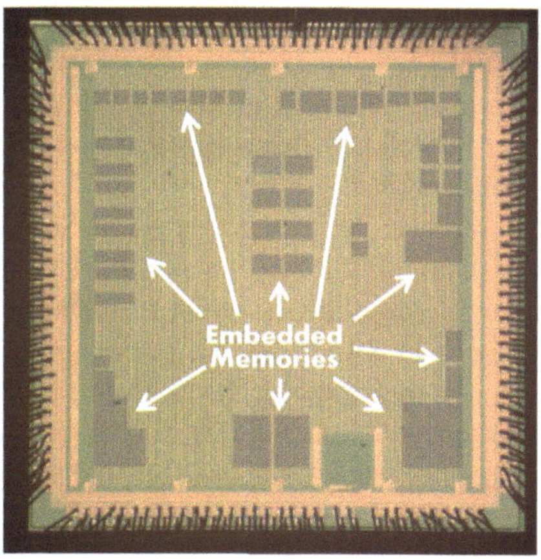

(c) Chip microphotograph of 4-stream 802.11n baseband transceiver [6].

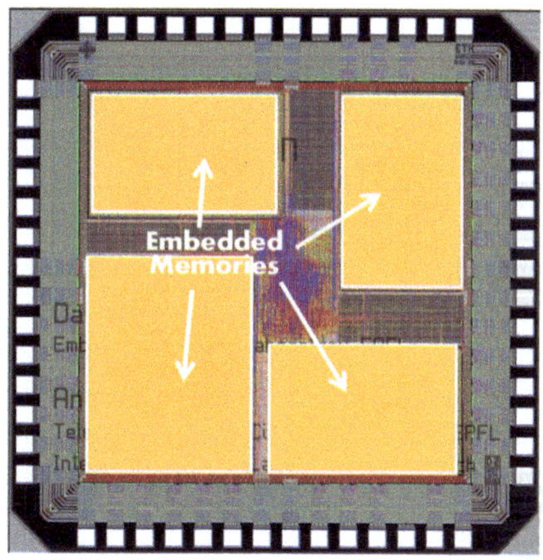

(d) Layout picture of an ultra-low power, sub-V_T microprocessor for biomedical applications.

Fig. 1.2 Layout pictures and/or chip microphotographs of high-end microprocessors (**a**–**b**), a baseband transceiver (**c**), and a low-power processor for biomedical signals (**d**). All these VLSI SoCs require a significant amount of embedded memories, which are visible as regular tiles in the layout

of the chip. Furthermore, the 4-stream 802.11n baseband transceiver [4], whose chip microphotograph is shown in Fig. 1.2c, contains a large number of SRAM macrocells which are visible as dark areas.

In addition to the large area share, embedded memories are also responsible for a large percentage of the power consumption of most VLSI SoCs. For example, the embedded memories of TamaRISC-CS, a ULP application-specific processor for compressed sensing [6], consume 70–95% of the total power, depending on the mode of operation. As a further example, in a configurable high-throughput decoder for quasi-cyclic low-density parity-check (LDPC) codes [23], the embedded memories are responsible for 68% of the total power consumption. Furthermore, as of today, VLSI SoCs for stationary applications typically have a total power consumption of up to 100 W, corresponding to the total of dynamic and static power consumptions of logic blocks and embedded memories [10], as shown in Fig. 1.3a. As opposed to this, Fig. 1.3b shows that VLSI SoC processors for portable applications have a considerably lower total power budget of 0.5 W, as per a requirement established by the ITRS in 2009. Only consumer processors for tablet computers may have a total power consumption as high as 2 W, given the physical product dimensions and advanced power management techniques [10]. For portable applications, the power consumption of embedded memories is expected to increase further and become almost 50% of the total power budget of processors in the next 15 years (see Fig. 1.3b). Reducing the power consumption of embedded memories is of utmost importance for all VLSI SoC application fields, for a number of quite different reasons. For example, low-power embedded memories and VLSI SoCs are essential to ensure runtimes of several years for ULP systems, such as implanted biomedical devices, to continue ensuring runtimes of ideally 1 day for portable computing devices of ever-increasing complexity (such as smartphones), or to reduce cooling costs for servers in data centers [8].

In addition to consuming dominant area and power percentages of VLSI SoCs, embedded memories are normally the first point of failure under voltage and technology down-scaling, due to the extremely high replication count of the basic bitcell, which, in most cases, is the 6-transistor (6T) SRAM bitcell. For example, if the supply voltage (V_{DD}) is scaled from its nominal value to the near-threshold (near-V_T) domain, the functional failure rate of embedded memories was shown to increase by five orders of magnitude [8]. As a consequence, under voltage and technology scaling, embedded memories typically limit the overall manufacturing yield of VLSI SoCs, whereas the complementary-metal-oxide-semiconductor (CMOS) based logic counterpart works more robustly.

1.2 Memory Requirements of Various Low-Power VLSI SoCs

Conventional personal computers and servers exhibit a deep memory hierarchy, ranging from on-chip, ultra-high speed, low storage capacity register files and cache memories, through fast, off-chip, higher capacity random-access memory (RAM), to slower, off-chip, high capacity, nonvolatile data storage. Traversing this memory hierarchy, the predominant, mainstream memory technologies are: (1) distributed or arrayed flip-flops or latches, (2) 6T-bitcell SRAM, (3) external, conventional 1-transistor-1-capacitor (1T-1C) dynamic random-access memory (DRAM), (4) Flash memory using a floating-gate transistor as a bitcell, and (5) mechanical hard disk drives, which are nowadays often replaced with solid-state drives. Note that only the register files and cache memories are embedded within the microprocessor chip, while the remaining parts of the computer memory hierarchy is off-chip.

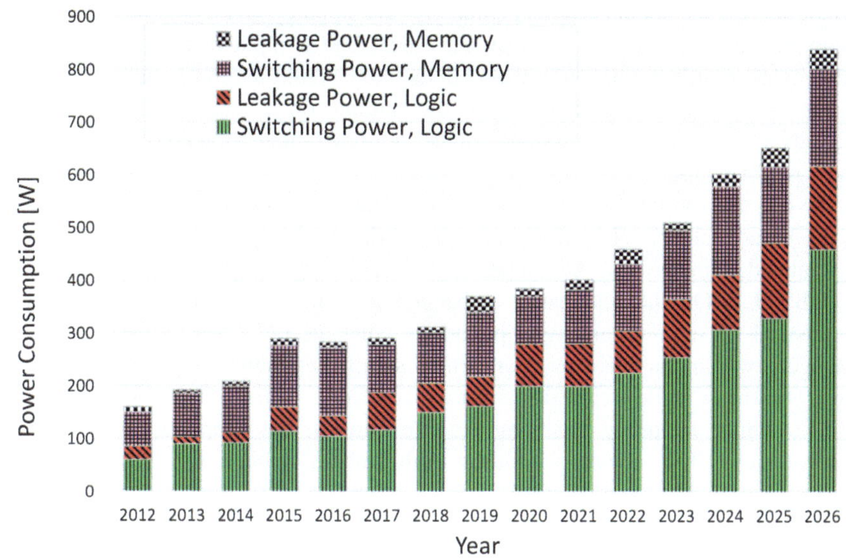

(a) Power breakdown of stationary consumer SoCs [2].

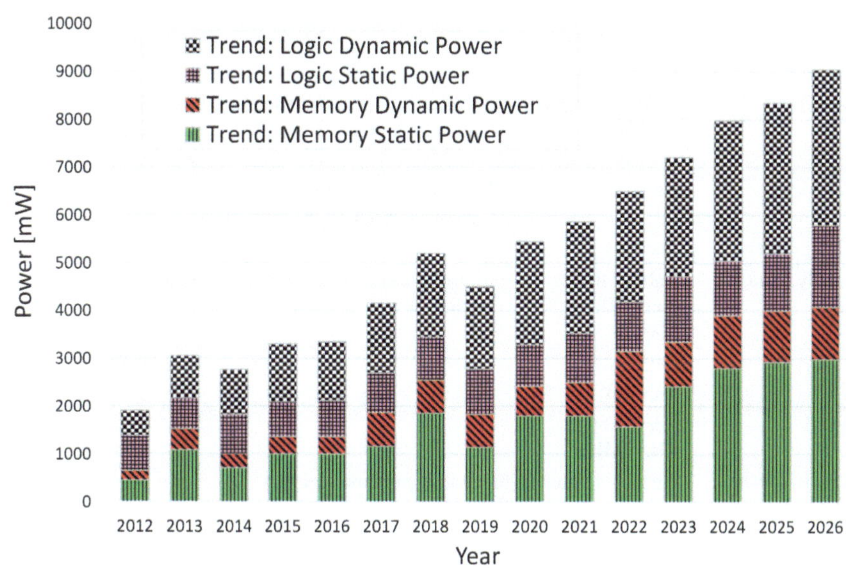

(b) Power breakdown of portable consumer SoCs [2].

Fig. 1.3 Predicted power breakdowns of VLSI SoCs for (**a**) stationary, and (**b**) portable consumer electronics [10]

Beside servers, personal computers, and laptop computers, battery-powered mobile computing devices such as smartphones and tablet computers impose extremely challenging requirements on embedded memory solutions due to the increasing power awareness—required to extend the runtime on a single battery charge—accompanied by an ever increasing demand for higher integration density and higher speed performance.

Beyond microprocessors for computers, a large number of target applications in the broad field of VLSI SoCs often have diametrically opposite requirements on embedded memories. A comparison of such target applications is provided in Table 1.1. On the one hand, embedded memories in ULP VLSI SoCs for biomedical or remote sensing applications (such as [27, 28]) require ultra-low leakage power and access energy and entail significant engineering effort to ensure high robustness, while area and speed are secondary concerns. Therefore, such ULP VLSI systems, including their embedded memories, are often operated at ultra-low voltages (ULV), typically residing in the sub-V_T domain. On the other hand, power-aware high-performance VLSI SoCs, often used in wireless communications

Table 1.1 Memory requirements of different classes of VLSI SoCs, from ultra-low power to power-aware, high-performance systems

	Ultra-low power	Low-power, medium-performance	Power-aware, high-performance
Application fields	Biomedical implants, remote sensors	Near-threshold computing, complex sensor nodes, simple handheld devices	Wireless communications, tablet computers, smartphones
Robustness	Robust	Potentially unreliable (detect + correct, or error-resilient)	
Area priority	Secondary		High
Supply voltage V_{DD}	Subthreshold (sub-V_T), e.g., 400 mV	Slightly scaled, near-threshold (near-V_T), e.g., 600 mV	Nominal, e.g., 1 V
Power	Ultra low, fW–pW		High, mW–W
Speed	Very slow, kHz–MHz		Fast, 100 MHz–GHz
State of the art	Bistables (latches, flip-flops), pipeline registers		
	8T, 10T, ..., 14T-bitcell SRAM, write and read assist	6T-bitcell SRAM, compilers 1T-1C eDRAM: special technology, extra cost Gain-cells: logic-compatible	
This book	Gain-cell eDRAMs (GC-eDRAMs)		
	2T sub-V_T	2T near-V_T	
	4T internal feedback	3T transmission-gate (TG)	
	4T redundant	Assist techniques: replica, body biasing	

(e.g., channel decoders) or in smartphones, require high-capacity, high-density, high-speed embedded memories operated at nominal supply voltages. In this case, rather than using robust, upsized SRAM bitcells, one-time programmable address decoders, if desired in combination with spare rows or columns to maintain storage capacity, are commonly used to cope with manufacturing defects (such as shorts and opens) [5]. Moreover, to cope with soft errors, caused by radiation occurring in the natural environment (e.g., alpha-particle impacts), redundant memory cells in conjunction with error detection and correction codes are often employed. A prominent example of such codes is the single-error-correction-double-error-detection (SECDED) code [11, 15]. Furthermore, as a new research direction, scientists and engineers have recently started to argue that the memory reliability can even be deliberately relaxed for VLSI systems which are inherently resilient to a small number of hardware defects. Examples of such inherently error-resilient systems include high-speed packet access (HSPA) systems [14] and wireless body sensor network (WBSN) nodes [20].

In addition to the above, an increasing number of VLSI systems feature dynamic voltage and frequency scaling (DVFS), in order to support different operating modes (such as high performance or low power modes), and/or reduce voltage and frequency guardbands for improved energy-efficiency and speed performance, respectively. Systems employing DVFS ideally contain embedded memories that are fully functional over the same voltage and frequency ranges as the logic. Besides the well-known Razor technique [9], as a further prominent example in the category of power-aware, high-performance VLSI SoCs supporting DVFS, Intel has presented an experimental, fully functional, error-resilient processor (codenamed Palisades) which has built-in mechanisms to detect and correct timing errors, allowing higher performance (by means of over-clocking) or better energy-efficiency (by means of voltage scaling) than a traditional processor with frequency and voltage guardbands [3].

In between the two extreme categories of ultra-low power VLSI SoCs operating in the sub-V_T domain and high-performance, power-aware, potentially error-resilient VLSI SoCs operating at nominal voltage, there is a third class corresponding to low-power, medium-performance SoCs (see Table 1.1). These SoCs and their embedded memories are typically operated at near-V_T supply voltages. Near-threshold computing (NTC) retains much of the energy savings of sub-V_T operation but has much better speed performance and suffers less from parametric variability [8]. An experimental, near-threshold voltage IA-32 microprocessor is able to successfully boot Windows XP™while being supplied from a small solar panel providing only 10–20 mW of power [1, 18]. As a further example of NTC SoCs, Diet SODA [21] is a power-efficient processor for digital cameras relying on near-threshold circuit operation.

1.3 Brief Review of the State of the Art

Broadly speaking, embedded memories can be divided into two main categories: (1) SRAM and (2) embedded DRAM (eDRAM). SRAM uses a cross-coupled inverter pair to retain the stored data statically as long as a power supply voltage is provided. The eDRAM technology stores data in the form of electric charge on a capacitor; unfortunately, the stored data is compromised due to leakage currents, which results in a requirement for a periodic refresh operation.

As shown in Table 1.1, latches and flip-flops (mostly implemented as static storage cells) are commonly used as pipeline registers or also in the form of small, synthesized storage arrays distributed within datapaths [26]. Static latches and flip-flops operate reliably at a large range of supply voltages, including sub-V_T voltages [16]. Memory macrocells based on the conventional 6T SRAM bitcell can be used for all applications running at nominal or slightly scaled supply voltages. In fact, almost invariably, SRAM has been the mainstream solution for on-chip embedded memories for virtually all VLSI SoC target applications for the last few decades [12]. This unquestioned dominance of SRAM technology for on-chip storage mostly arises from their fast write and read accesses and their robust operation in mature CMOS nodes and at nominal supply voltages. Also, for most process nodes, SRAM memory compilers are readily available, facilitating their wide deployment. However, the footprint of the 6T SRAM bitcell is relatively large, since six transistors need to be accommodated. In order to increase the storage density, eDRAM macrocells are an interesting alternative to SRAM macrocells. We distinguish between two types of eDRAM: (1) conventional, one-transistor, one-capacitor (1T-1C) eDRAMs, whose basic bitcell is built from a special, high-density, 3D capacitor and a single access transistor, and (2) gain-cell eDRAMs (GC-eDRAMs) (e.g., [22]), whose basic bitcell is built from 2 to 4 MOS transistors [25]. Conventional 1T-1C eDRAMs typically require special process options to build high-density stacked or trench capacitors [13] and are therefore not compatible with the widespread standard digital CMOS technology. Such process options are only available at an extra manufacturing cost and are not readily available for all technology processes. As opposed to this, GC-eDRAMs are fully compatible with baseline digital CMOS technologies and can easily be integrated into any SoC at no extra cost. The main drawback of gain-cells is the small storage node capacitor (compared to the dedicated DRAM capacitors) and the resulting low retention time. From a functional perspective, all types of dynamic memories usually require data refresh cycles which are costly in terms of power and have a small access bandwidth penalty.

6T-bitcell SRAM fails to operate reliably at aggressively scaled supply voltages [24]. As shown in Table 1.1, alternative SRAM bitcells consisting of 8, 10, or even up to 14 transistors are required to ensure reliable sub-V_T operation [19]. In addition to large, alternative SRAM bitcells, various low-voltage write and read assist techniques have been proposed.

1.4 Book Outline

As shown in Table 1.1, this book presents recent research results from the field of gain-cell eDRAM design. Various GC-eDRAMs are specifically designed and optimized for a range of low-power VLSI SoC applications, ranging from power-aware high-performance, through low-power medium-performance, to ultra-low power SoCs.

After a general introduction to embedded memories, given earlier in this chapter, Chap. 2 explains the basic operating principle of GC-eDRAMs, and discusses their advantages and drawbacks compared to SRAM and conventional 1T-1C eDRAM. A detailed review of prior-art GC-eDRAM circuit techniques and target applications is provided next, unveiling that most previous GC-eDRAM works focused on high-performance processor applications, while the field of low-power low-voltage GC-eDRAM design was mostly unexplored before the publication of this book.

Since GC-eDRAMs have a small storage node capacitor compared to conventional 1T-1C eDRAMs, their retention time is limited, and data refresh cycles are responsible for a large power share. In order to optimize GC-eDRAMs for low-power VLSI SoCs, it is thus key to first develop a deep understanding of the factors limiting the retention time, and then develop techniques to extend the retention time. To this end, Chap. 3 introduces an analytical retention time model, and further develops this analytical model to describe the statistical retention time distribution arising from parametric variations. The high accuracy of the analytical model is verified by means of circuit simulations and the manufacturing and measurement of GC-eDRAM test arrays. To summarize the learnings from detailed retention time sensitivity analyses, best-practice 2-transistor (2T) gain-cell (GC) design guidelines are presented.

Supply voltage scaling to the near-V_T domain is a prominent low-power circuit technique often used to minimize the power consumption of logic VLSI circuits. Chapter 4 argues that embedded memories should follow this trend in order to facilitate SoC integration. In this context, the impact of supply voltage scaling on the retention time of a conventional 2T GC-eDRAM array is analyzed, and it is further shown that the retention time depends on the write access statistics. In a retention scenario where the array is not updated after an initial full array write cycle, the write bit-line (WBL) can be controlled. The chapter concludes that thanks to WBL control during retention modes, voltage scaling can improve the retention time. In summary, Chap. 4 shows that a conventional 2T GC bitcell and array organization can be operated at a near-V_T voltage, and that a WBL control scheme during retention modes can improve the retention time under voltage scaling.

After reviewing conventional 2T GC-eDRAM operated at near-V_T voltages, Chap. 5 presents novel GC-eDRAM bitcells and assist techniques specifically designed to enhance the data retention time for operation at near-V_T. First, a 3-transistor (3T) GC using a full transmission gate (TG) write port is presented. This full TG 3T GC bitcell allows fast write and memory operation at a single supply voltage, whereas conventional 2T GC-eDRAMs require a write word-line (WWL) boost in addition to the main supply voltage. Furthermore, the full TG 3T GC bitcell enables strong initial data levels on the storage node (SN) for extended retention times. Next, the impact of body biasing on a 2T GC-eDRAM is analyzed. Silicon measurements of a test array implemented in 0.18 μm CMOS technology show that reverse body biasing (RBB) can significantly extend the retention time. Finally, as an assist technique to reduce accumulated pessimism from assuming worst-case process, voltage, temperature (PVT) conditions and write disturb activities, a replica technique for optimum refresh timing is presented, and its effectiveness is demonstrated by silicon measurements.

In order to study the suitability of GC-eDRAM beyond mature technology nodes such as 0.18 μm CMOS and beyond near-V_T supply voltages, Chap. 6 analyzes aggressive technology scaling, down to 28 nm CMOS, as well as aggressive supply voltage scaling, all the way to the sub-V_T domain. First, the analytical retention time model developed in Chap. 3 is validated for a scaled 28 nm CMOS node. Second, a conventional 2T GC topology is optimized for sub-V_T operation. Simulations show successful sub-V_T operation for a 2T GC-eDRAM array in 0.18 μm CMOS. Furthermore, a 2T GC-eDRAM array implemented in a scaled 40 nm CMOS technology can be operated successfully down to the near-V_T domain, while simultaneous aggressive technology and voltage scaling to the sub-V_T domain are not recommended.

After evaluating conventional 2T GC-eDRAMs in deeply scaled CMOS nodes and in the sub-V_T regime, Chap. 7 presents novel GC-eDRAM bitcells specifically designed and optimized for the most advanced CMOS nodes, as well as for soft error tolerance which becomes an increasingly important issue with technology scaling. A novel 4-transistor (4T) GC with selective internal feedback to protect only the weaker data level among "0" and "1" is presented. A simulation based proof of concept is provided for a 65 nm CMOS node. A redundant 4T GC bitcell for soft error tolerance is presented next. This 4T GC offers per-cell redundancy at a small area cost and enables GC-eDRAM array architectures with a parity column for error correction.

Chapter 8 concludes the book.

References

1. A solar powered IA core? no way! Intel Developer Forum (2011). http://blogs.intel.com/research/2011/09/ntvp
2. Bohr M (2009) The new era of scaling in an SoC world. In: IEEE international solid-state circuits conference (ISSCC), pp 23–28
3. Bowman K, Tschanz J, Lu S, Aseron P, Khellah M, Raychowdhury A, Geuskens B, Tokunaga C, Wilkerson C, Karnik T, De V (2011) A 45 nm resilient microprocessor core for dynamic variation tolerance. IEEE J Solid State Circuits 46(1):194–208
4. Burg A, Haene S, Borgmann M, Baum D, Thaler T, Carbognani F, Zwicky S, Barbero L, Senning C, Greisen P, Peter T, Foelmli C, Schuster U, Tejera P, Staudacher A (2009) A 4-stream 802.11n baseband transceiver in 0.13 μm CMOS. In: IEEE symposium on VLSI circuits, pp 282–283

5. Bushnell M, Agrawal V (2000) Essentials of electronic testing for digital, memory and mixed-signal VLSI circuits, chap 9.1. Springer, Berlin
6. Constantin J, Dogan A, Andersson O, Meinerzhagen P, Rodrigues J, Atienza D, Burg A (2012) TamaRISC-CS: An ultra-low-power application-specific processor for compressed sensing. In: Proceedings of the IEEE/IFIP international conference on VLSI system-on-chip (VLSI-SoC), pp 159–164
7. Damaraju S, George V, Jahagirdar S, Khondker T, Milstrey R, Sarkar S, Siers S, Stolero I, Subbiah A (2012) A 22 nm IA multi-CPU and GPU system-on-chip. In: IEEE international solid-state circuits conference (ISSCC), pp 56–57
8. Dreslinski R, Wieckowski M, Blaauw D, Sylvester D, Mudge T (2010) Near-threshold computing: reclaiming moore's law through energy efficient integrated circuits. Proc IEEE 98(2):253–266
9. Ernst D, Kim NS, Das S, Pant S, Rao R, Pham T, Ziesler C, Blaauw D, Austin T, Flautner K, Mudge T (2003) Razor: a low-power pipeline based on circuit-level timing speculation. In: Proc. IEEE/ACM international symposium on microarchitecture (MICRO), pp 7–18
10. International technology roadmap for semiconductors (2011). http://www.itrs.net/Links/2011ITRS/Home2011.htm
11. Jahinuzzaman S, Shah J, Rennie D, Sachdev M (2009) Design and analysis of a 5.3-pJ 64-kb gated ground SRAM with multiword ECC. IEEE J Solid State Circuits 44(9):2543–2553
12. Kaeslin H (2008) Digital integrated circuit design: from VLSI architectures to CMOS fabrication, 1st edn. Cambridge University Press, Cambridge
13. Kang S, Leblebici Y (2003) CMOS digital integrated circuits: analysis and design, 3rd edn. McGraw-Hill, New York
14. Karakonstantis G, Roth C, Benkeser C, Burg A (2012) On the exploitation of the inherent error resilience of wireless systems under unreliable silicon. In: Proceedings of the ACM/EDAC/IEEE design automation conference (DAC), pp 510–515
15. Kumar A, Rabaey J, Ramchandran K (2009) SRAM supply voltage scaling: a reliability perspective. In: Proceedings of the IEEE international symposium on quality electronic design (ISQED), pp 782–787
16. Meinerzhagen P, Sherazi S, Burg A, Rodrigues J (2011) Benchmarking of standard-cell based memories in the sub-VT domain in 65-nm CMOS technology. IEEE J Emerging Sel Top Circuits Syst 1(2):173–182
17. Nehalem part 3: The cache debate, LGA-1156 and the 32 nm future (2008). http://www.anandtech.com/show/2671
18. Pande PP, Ganguly A, Chakrabarty K (2013) Chapter 9: Claremont: a solar-powered near-threshold voltage IA-32 processor. In: Vangal S, Jain S (eds) Design technologies for green and sustainable computing systems. Springer, Berlin
19. Qazi M, Sinangil M, Chandrakasan A (2011) Challenges and directions for low-voltage SRAM. IEEE Des Test Comput 28(1):32–43
20. Sabry MM, Karakonstantis G, Atienza D, Burg A (2012) Design of energy efficient and dependable health monitoring systems under unreliable nanometer technologies. In: Proceedings of the ACM international conference on body area networks, pp 52–58
21. Seo S, Dreslinski R, Woh M, Chakrabarti C, Mahlke S, Mudge T (2010) Diet SODA: a power-efficient processor for digital cameras. In: Proceedings of the ACM/IEEE international symposium on low-power electronics and design (ISLPED), pp 79–84
22. Somasekhar D, Ye Y, Aseron P, Lu SL, Khellah M, Howard J, Ruhl G, Karnik T, Borkar S, De V, Keshavarzi A (2008) 2GHz 2Mb 2T gain-cell memory macro with 128GB/s bandwidth in a 65nm logic process. In: Proceedings of the IEEE international solid-state circuits conference (ISSCC), pp 274–613
23. Studer C, Preyss N, Roth C, Burg A (2008) Configurable high-throughput decoder architecture for quasi-cyclic LDPC codes. In: Proceedings of the IEEE Asilomar conference on signals, systems and computers, pp 1137–1142
24. Teman A, Visotsky R (2015) A fast modular method for true variation-aware separatrix tracing in nanoscaled SRAMs. IEEE Trans Very Large Scale Integr 23(10):2034–2042

25. Teman A, Meinerzhagen P, Burg A, Fish A (2012) Review and classification of gain cell eDRAM implementations. In: Proceedings of the IEEE convention of electrical and electronics engineers in Israel (IEEEI), pp 1–5
26. Teman A, Rossi D, Meinerzhagen P, Benini L, Burg A (2016) Power, area, and performance optimization of standard cell memory arrays through controlled placement. ACM Trans Des Autom Electron Syst (TODAES) 21(4):59
27. Yoo J, Yan L, El-Damak D, Bin Altaf M, Shoeb A, Yoo HJ, Chandrakasan A (2012) An 8-channel scalable EEG acquisition SoC with fully integrated patient-specific seizure classification and recording processor. In: Proceedings of the IEEE international solid-state circuits conference (ISSCC), pp 292–294
28. Zhang F, Zhang Y, Silver J, Shakhsheer Y, Nagaraju M, Klinefelter A, Pandey J, Boley J, Carlson E, Shrivastava A, Otis B, Calhoun B (2012) A batteryless 19 μ W MICS/ISM-band energy harvesting body area sensor node SoC. In: Proceedings of the IEEE international solid-state circuits conference (ISSCC), pp 298–300

Chapter 2
Gain-Cell eDRAMs (GC-eDRAMs): Review of Basics and Prior Art

2.1 Basics of GC-eDRAM

While 6-transistor (6T)-bitcell static random-access memory (SRAM) macrocells are the mainstream solution for embedded memories in very large scale integration (VLSI) systems-on-chip (SoCs), as discussed in Chap. 1, embedded dynamic random-access memory (eDRAM) macrocells are an interesting area-efficient alternative. The conventional eDRAM bitcell uses a dedicated storage capacitor to store information in the form of electric charge. It further uses a single metal-oxide-semiconductor field-effect transistor (MOSFET) to access the storage capacitor for read and write operations. Unfortunately, such conventional 1-transistor-1-capacitor (1T-1C)-bitcell eDRAM requires special processing steps to manufacture high-density stacked or trench capacitors. It is therefore not directly compatible with standard digital complementary metal-oxide-semiconductor (CMOS) technologies [16] and not readily available for integration with logic in all process flavors and nodes.

As opposed to conventional 1T-1C eDRAM, gain-cell (GC) based eDRAM (GC-eDRAM) is fully compatible with mainstream digital CMOS technologies, since it is built exclusively from MOSFETs and, optionally, the readily available metal stack and vias. MOSFETs are used as access transistors and as MOSCAPs. Metal layers and vias can be used to enhance the storage node capacitance. As such, GC-eDRAM is an interesting alternative to 6T-bitcell SRAM and 1T-1C eDRAM, since it combines many of the advantages of SRAM (e.g., the compatibility with digital CMOS technologies) and 1T-1C eDRAM (e.g., higher storage density than SRAM), while it avoids most of the drawbacks of SRAM (e.g., the large bitcell) and of 1T-1C eDRAM (e.g., the destructive read, write-back operation, and extra cost for special process options). Thanks to its compatibility with standard digital CMOS technologies, GC-eDRAM macrocells can readily be integrated with any digital system at no additional manufacturing cost for special process options, as opposed

P. Meinerzhagen et al., *Gain-Cell Embedded DRAMs for Low-Power VLSI Systems-on-Chip*, DOI 10.1007/978-3-319-60402-2_2

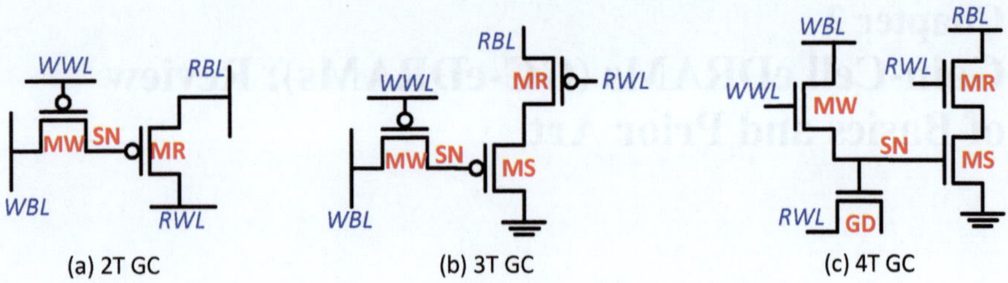

Fig. 2.1 Basic 2–4T gain-cells (GCs). (**a**) 2T GC. (**b**) 3T GC. (**c**) 4T GC

to 1T-1C eDRAM. The main drawback of GC-eDRAM compared to SRAM is the dynamic characteristic of the data retention, which usually requires periodic refresh operations.

A large variety of different GC topologies have been proposed in the last decade, consisting of 2–4 transistors. Basic examples of 2T, 3T, and 4T GCs are shown in Fig. 2.1. All of these circuits exhibit a write transistor (MW) to access the capacitive storage node (SN) and deposit charge on it. Moreover, all GC topologies have an SN capacitor which consists of a dedicated MOSCAP, the junction capacitance of MW, and in some cases of sidewall and parallel-plate capacitors built above the cell footprint with the available metal and via resources. In the smallest 2-transistor (2T) GC configuration (see Fig. 2.1a), the dedicated storage transistor, i.e., the MOSCAP, is also used as read transistor (MR). The slightly larger 3-transistor (3T) GC configuration (see Fig. 2.1b) exhibits a more robust read operation by using a separate MR, which decouples the read bitline (RBL) from MR. Some 4-transistor (4T) GCs (see Fig. 2.1c) use an additional MOSCAP to increase the SN capacitor and to capacitively couple the RBL to the SN for increased read robustness. The term "gain-cell" stems from the transconduction gain of the read transistor MR, which translates a voltage level on the SN, or, equivalently, the gate voltage of MR, into an output sense current, i.e., the drain current of MR. From a similar point of view, the term "gain" can also relate to the fact that a small amount of charge on the SN leads to a large charge flow on the RBL during readout thanks to the use of MR [21].

2.2 Advantages and Drawbacks of GC-eDRAM

GC-eDRAM has several advantages compared to both SRAM and 1T-1C eDRAM. In fact, a GC is significantly smaller than a 6T SRAM bitcell; typically, area savings of at least 50% can be achieved by employing GCs instead of SRAM bitcells. Moreover, GCs have much lower aggregated bitcell leakage currents than SRAM bitcells. This reduced bitcell leakage current can even lead to lower GC-eDRAM data retention power, i.e., the sum of leakage and active refresh power, compared to the static leakage power of a corresponding SRAM macrocell [6]. Compared to conventional 1T-1C eDRAM, GC-eDRAM does not require any special processing steps to build high-density trench or stacked capacitors [16], which would require 4–6 extra masks and would add cost to a digital CMOS process [14]. As a further advantage compared to 1T-1C eDRAM, GCs enable a nondestructive read operation and thereby avoid the need for a write-back or restore operation.

Furthermore, compared to both the 6T SRAM bitcell and the 1T-1C bitcell, all GC topologies have a separate read and write port, which allows the construction of two-port GC-eDRAM macrocells at virtually no area overhead compared to single-port macrocells. As opposed to that, both the 6T SRAM bitcell and the 1T-1C bitcell share the same bit-line(s) (BL) and word-line (WL) for both write and read accesses. Therefore, additional hardware is required in each basic storage cell to allow simultaneous write and read access to a storage array built from SRAM or conventional DRAM cells. The use of two-port GC-eDRAM macrocells is appealing to ensure high memory bandwidth compared to single-port macrocells [15]. This can be especially interesting to recover some of the speed penalty resulting from voltage scaling, which is required for low power consumption, or simply to ensure high access bandwidth for GC-eDRAMs used as caches in high-performance microprocessors. Finally, the separate write and read ports of all GC topologies allow to independently and simultaneously optimize the bitcell for enhanced write-ability and read-ability. This is especially important for the implementation of embedded memories in aggressively scaled CMOS nodes, characterized by high parametric variations, and/or operated at low voltages, in which case parametric variations become problematic due to degraded on/off current ratios. Note that the possibility of simultaneously and independently sizing the transistors in a GC for robust read and robust write is a unique property of GCs which cannot be found in the 6T SRAM bitcell or in the 1T-1C eDRAM cell. In fact, in case of SRAM bitcells, additional transistors are required to avoid write contention and to improve read-ability.

These various advantages of GCs compared to the traditional 6T SRAM and 1T-1C bitcells motivate the analysis and optimization of GC-eDRAM for use as embedded memories in a large variety of future low-power VLSI SoCs implemented in scaled CMOS nodes and operated a scaled voltages.

Beside this long list of advantages, the main drawback of GC-eDRAM, compared to SRAM, is the dynamic storage mechanism, which requires periodic, power-consuming refresh cycles (unless the memory block is anyway periodically updated, such as the internal memories of the LDPC decoder presented in [27]). Compared to the conventional 1T-1C eDRAM bitcell, the total in-cell storage capacitor of GCs is considerably smaller, which leads to shorter retention times and requires more frequent refresh cycles. Also, there is a large variability of per-cell retention time across a GC-eDRAM array [7, 25], and, unfortunately, the global refresh rate needs to be set according to the GC with the worst retention time, unless spare rows or columns in conjunction with programmable address decoders are used [1]. Later on in this book, in Chaps. 4 through 7, we present several techniques to improve the retention time of GC-eDRAM in order to render it even more attractive for use in future low-power VLSI SoCs. However, before presenting our novel GC-eDRAM designs, a detailed review of the field of GC-eDRAM is presented in the remainder of this chapter, which also positions the herein presented work with respect to prior-art GC-eDRAM implementations.

2.3 Review of Prior-Art GC-eDRAM Circuit Techniques and Target Applications

2.3.1 Categorization of GC-eDRAM Implementations

From the large amount of recent publications on GC-eDRAM, it is possible to identify four main categories of target applications: (1) high-end processors requiring large embedded cache memories; (2) general system-on-chip designs; (3) low-voltage low-power systems, such as biomedical systems; and (4) fault-tolerant systems including channel decoders for wireless communications. The following sections provide more details on each of these four categories.

2.3.1.1 Gain-Cells for High-End Processors

The vast majority of recent research on GC-eDRAM is dedicated to large embedded cache memories for microprocessors [3–6, 20–22, 29–32]. In fact, GC memories are considered to be an interesting alternative to SRAM, which has been the dominant solution for cache memories for several decades. This is due to the higher density, increased speed, and potentially lower leakage power of GC-eDRAM w.r.t. SRAM. Besides the obvious advantage of high integration density, the main design goals for GC-eDRAMs in this application category are high speed operation and high memory bandwidth, especially for industrial players like IBM [22] and Intel [30, 31], and more recently also for academia [4, 6]. A smaller number of research groups specify low power consumption as their primary design goal [3, 5]. In fact, as mentioned before, a recent study shows that GC-eDRAMs potentially consume less data retention power (sum of leakage and refresh power) than SRAM arrays (leakage power only) [6].

2.3.1.2 General Systems-on-Chip (SoCs)

Several authors are not very specific about their target applications [2, 11, 12], as they only mention general SoCs. However, they follow the same trend as the aforementioned high-end processor community by proposing GC-eDRAMs as a replacement for the mainstream 6T-bitcell SRAM solution. For these SoC applications, the main drivers are the potential for higher density and lower power consumption than SRAM.

2.3.1.3 Gain-Cells for Ultra-Low Power (Biomedical) Systems

While the previously described target applications require relatively high memory bandwidth, several recent GC-eDRAM publications target low-voltage low-power applications, mostly in the biomedical domain. After a short preview in this section, GC-eDRAM design for low-voltage low-power applications will be extensively discussed throughout the remainder of this book. A GC-eDRAM implemented in a mature low-leakage 180 nm CMOS process achieves low retention power through voltage scaling well below the nominal supply voltage [19]. The positive impact of supply voltage scaling on retention time for given access statistics and a given write bit-line (WBL) control scheme is demonstrated in [13] and expatiated on in Chap. 4, proposing near-threshold (near-V_T) operation for long retention times and therefore low retention power. Moreover, a full transmission gate (TG) write port [10], reverse body biasing (RBB) [25], and replica techniques have been proposed in order to further enhance the retention time and reduce the power consumption of near-V_T GC-eDRAM macrocells, as will be shown in Chap. 5. Furthermore, recent studies [24, 26] show that the supply voltage of GC arrays can even be scaled down to the subthreshold (sub-V_T) domain, while still guaranteeing robust operation and high memory availability for read and write operations; more details on these studies are given in Chap. 6. Finally, partial internal feedback is used in [8] in order to increase the data retention time and achieve ultra-low retention power, while area-efficient per-cell redundancy is proposed in [9] for increased soft error tolerance of ultra-low power SoCs; details on these techniques are provided in Chap. 7.

2.3.1.4 Gain-Cells for Wireless Communications Systems

A small number of recently presented GC-eDRAM designs are fundamentally different from the aforementioned works, as they are specifically built and optimized for systems which require only short retention times, and in some cases, are tolerant to a small number of hardware defects (such as read failures) [17]. The refresh-free GC-eDRAM used in a recently published low-density parity-check (LDPC) decoder is periodically updated with new data, and therefore requires a retention time of only 20 ns [28]. Besides safely skipping power-hungry refresh cycles and designing for low retention times, a few works [18, 23] also exploit the fact that wireless communications systems and other fault-tolerant systems are inherently resilient to a small number of hardware defects. In fact, by proposing memories based on multilevel GCs, the storage density of GC-eDRAMs is further increased at the price of a small number of read failures which do not significantly impede the system performance [18, 23].

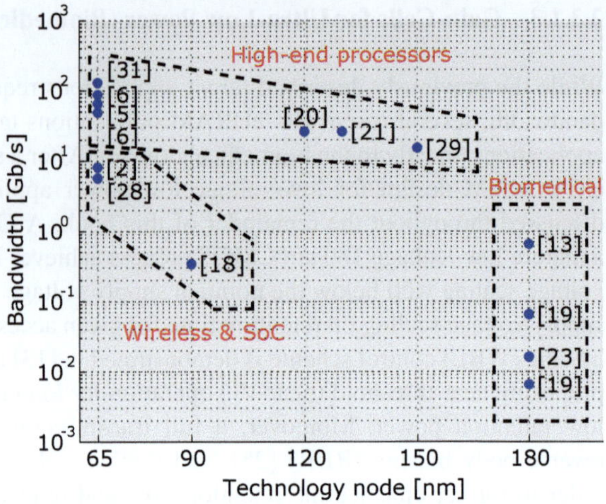

Fig. 2.2 Bandwidth vs. technology node of several published GC-eDRAM implementations

2.3.2 Comparison of the State-of-the-Art Implementations

Figure 2.2 shows the bandwidth and the technology node of the state-of-the-art GC-eDRAM implementations, highlighted according to target application categories. References appearing multiple times correspond to different operating modes or operating points of the same design. The figure shows a difference of more than four orders-of-magnitude in the achieved memory bandwidth among the various implementations. GC-eDRAMs designed as cache memory for processors achieve around 10 Gbit/s, if implemented in older technologies, and over 100 Gbit/s if implemented in a more advanced 65 nm CMOS node. Most memories designed for wireless communications systems or generally for SoCs still achieve bandwidths between 1 and 10 Gbit/s. Only the high-density multilevel GC array has a lower bandwidth due to a slow successive approximation multilevel read operation [18]. GC-eDRAMs targeted at biomedical systems are preferably implemented in a mature, reliable 180 nm CMOS node and achieve sufficiently high bandwidths between 10 Mbit/s and several 100 Mbit/s at near-V_T or sub-V_T supply voltages.

Figure 2.3 shows the retention power, i.e., the sum of refresh and leakage power, of previously reported GC-eDRAMs versus their retention time. For energy-constrained biomedical systems, long retention times of 1–10 ms are a key design goal in order to achieve low retention power between 600 fW/bit and 10 pW/bit. The memory banks of the LDPC decoder in [28] have a nominal retention time of 1.6 µs, which is around four orders-of-magnitude lower than that of the arrays targeted at biomedical systems. Even though the reported power consumption of 5 µW/bit corresponds to active power [28], it is fair to compare it to the retention power of other implementations, as data would anyway need to be refreshed at the same rate as new data is written. Interestingly, the power consumption per bit of this refresh-free eDRAM is almost seven orders-of-magnitude higher than the retention power per bit of the most efficient eDRAM implementation for biomedical systems. The retention time and retention power of GC-eDRAMs for processors are in between

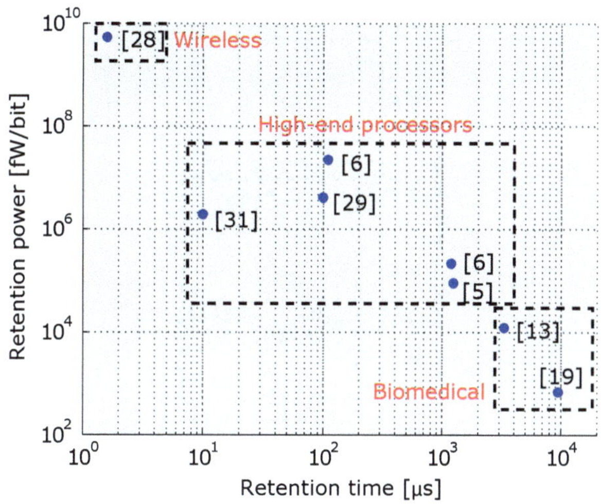

Fig. 2.3 Retention power vs. retention time for several published GC-eDRAM implementations

the values for the wireless and biomedical application domains. Overall, of course, it is clearly visible that enhancing the retention time is an efficient way to lower the retention power.

The *area cost per bit* (ACPB) is defined as the silicon area of the entire memory macro (including peripheral circuits), divided by the storage capacity. As opposed to the simple bitcell size metric, ACPB accounts for the area overhead of peripheral circuits and is a more suitable metric to compare different memory implementations. Moreover, we define the *array efficiency* as the bitcell size divided by the ACPB; note that the array efficiency is a technology-independent metric. Figure 2.4 shows the comparably higher ACPB of biomedical GC-eDRAMs due to the use of a mature 180 nm CMOS node. However, despite their small storage capacity requirements, these implementations achieve a high array efficiency of over 0.5, by using small, yet slow, peripheral circuits [19]. As opposed to this, all GC-eDRAMs targeted at processors, wireless communications, and SoC applications achieve array efficiencies below 0.5, meaning that over half of the area of those macrocells is occupied by peripheral circuits.

2.3.3 *Circuit Techniques for Target Applications*

GC-eDRAMs have been shown to be an attractive alternative to traditional SRAM arrays for large caches in high-end processors, wireless communication systems, and ultra-low power VLSI systems. Hereinafter, the circuit techniques used in these GC-eDRAM implementations are presented and related to their respective target metrics.

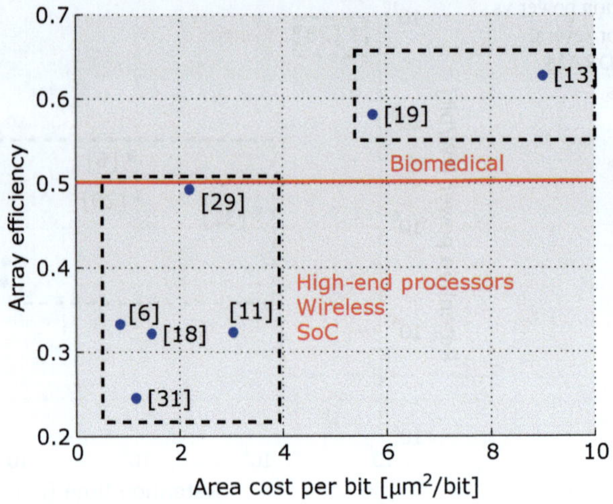

Fig. 2.4 Array efficiency vs. area cost per bit (ACPB) for several published GC-eDRAM implementations

2.3.3.1 Gain-Cell Topologies

An extensive comparison between recent GC topologies is presented in Table 2.1. A common feature of all these circuits is their reduced transistor count, as compared to traditional SRAM circuits. The highest device count appears in [22], comprising three transistors and a "gated diode" (MOS transistor acting as storage device and amplifier for the cell voltage), with all other proposals made up of three [2, 3, 5, 11, 18, 20, 21, 23, 28] or two [6, 13, 19, 24, 26, 29–31] transistors. The obvious implication of the transistor count is the bitcell area; however, the choice of the topology is application dependent, as well. The simple structure of the 2T topologies usually includes a write transistor (MW) and a combined storage and read transistor (MR). Transistor MW connects the write bit-line (WBL) to the storage node (SN) when the write word-line (WWL) is asserted, and transistor MR amplifies the stored charge signal by driving a current through the read bit-line (RBL) when the read word-line (RWL) is asserted. The 2T structure results in capacitive coupling effects between the control lines (WWL, RWL, RBL) and the SN, which can affect the data integrity and degrade performance. Therefore, a third device is often added, primarily to avoid disturbing capacitive couplings from the RWL onto the SN and to reduce RBL leakage. These 3T GC configurations give up some of the density advantage of 2T GCs for the benefit of enhanced speed performance, robustness, and/or retention time. The boosted 3T topology of [5] utilizes the capacitive coupling effect to extend the retention time by connecting the drain of MR to RWL, rather than ground, thereby negating some of the positive SN voltage step inherent to the PMOS MW configurations. Interestingly, large cache memory designs [6, 29, 31] prefer the 2T topology at the cost of additional peripheral circuits to retain high speed performance. An interesting choice of the

Table 2.1 Overview of gain-cell circuit techniques according to target applications

Category	High Performance Processor Caches					
Publication	[20, 21, 2]	[29]	[22]	[30, 31]	[3, 5]	[6]
Bitcell	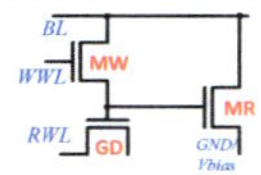	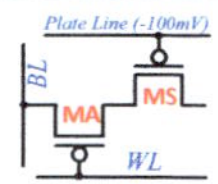	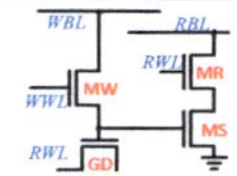	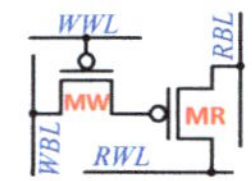	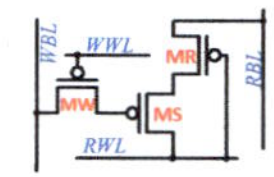	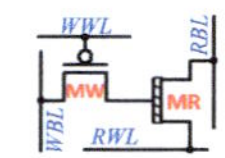
Tech. Node	0.12 µm, 0.13 µm, 65 nm PTM	0.15 µm	90 nm	65 nm	65 nm	65 nm
Techniques	Gated Diode, Footer Power Gating, Foot Driver	Multi-Level Bitlines, Hybrid open bitline architecture	Gated Diode Sense Amplifier	RBL Clamping, Pipelined Architecture	Boosted 3T, PVT tracking read reference feedback, Regulated WBL	Half Swing WBL, Stepped WWL
Main Design Metric	400 MHz, 70 µs retention, 100 kbit	400 MHz, 100 µs retention, 1 Mbit	up to 2 GHz, 110 µs retention, 40 kbit	2 GHz, 10 µs retention, 2 Mbit	500 MHz, up to 1.25 ms ret., 64 kbit	667 MHz, 110 µs ret., 192 kbit
Category	General SoC		Wireless	Low Power Biomedical Systems		
Publication	[11]	[27, 18]	[28]	[19]	[13]	[23, 25]
Bitcell	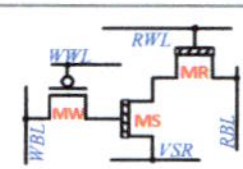	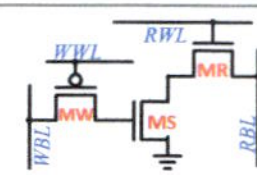	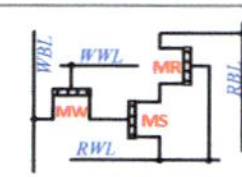	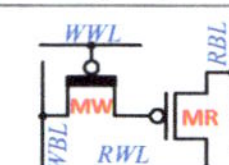	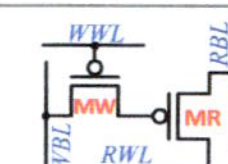	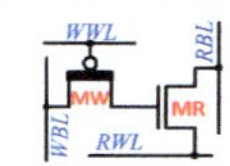
Tech. Node	90 nm	90 nm	65 nm	0.18 µm	0.18 µm	0.18 µm
Techniques	Forced Feedback, Write Echo Refresh	Multi Level Bitcell, PVT Replica Column	Refresh Free, Sequential Decoding	I/O Write Transistor, Low Area Sense Buffer	Low Area Sense Buffer	Hybrid Cell with I/O MW, Sense Buffer
Main Design Metric	V_{DD} =0.5 V, 180 µA ref. power, 5 MHz	2–50 µs retention, 1.45 µm ^{2}/bit density	32× 1 kbit arrays, 700 MHz, 170 ns retention	V_{DD}=0.75 V, up to 306 ms ret., 0.1–1 MHz, 662 fW/bit ret. power	V_{DD}=0.75 V, 3.3 ms retention, 11.9 pW/bit ret. power	V_{DD}=400 mV, over 40 ms ret., 500 kHz

2T topology is used in [24] even though the target application is a small array for ultra-low power (biomedical) systems. In this case, the stacked readout path of the 3T topology proved to be too slow under sub-V_T supply voltages.

2.3.3.2 Device Choices

The majority of today's CMOS process technologies provide several device choices, manipulating the oxide thickness and channel implants to create several threshold voltage (V_T) and maximum voltage tolerance options. Careful choice of the appropriate device (PMOS/NMOS, standard/high/low V_T) can provide orders-of-magnitude improvement in GC performance, as apparent in Table 2.1. PMOS devices typically suffer from lower drive strength than their NMOS counterparts, but have substantially lower subthreshold and gate leakage currents. For most process technologies, the primary cause of storage node charge loss is subthreshold leakage current through MW, and therefore the ultra-low power implementations [19, 24] employ a high-V_T or I/O PMOS to substantially extend the retention time. In addition to subthreshold leakage, gate leakage current is substantial in thin oxide CMOS technologies. Therefore, the all-PMOS 2T configuration [31] balances the subthreshold and gate leakage currents out of and in to the SN to improve retention time. The decoder system of [28] requires high speed performance with very short retention times, and therefore, an all NMOS low-V_T circuit is used. Low-V_T devices are used in the readout path of several other publications [6, 11] in order to improve the read speed without increasing the static power, since there is a zero drain-to-source voltage drop across MR during write and standby cycles.

The device choices affect the capacitive couplings to and charge injection onto the SN. The MW device type significantly affects the initial voltage level of the SN, depending on several factors. A PMOS write transistor passes a weak "0," and an NMOS passes a weak "1"; therefore, an underdriven (for PMOS MW) or boosted (for NMOS MW) WWL voltage is necessary to pass a full voltage level to the SN. However, the larger the WWL voltage swing is, the larger the capacitively coupled voltage step on the storage node during WWL de-assertion. A PMOS MW is cut-off by the rising edge of WWL, resulting in both capacitive coupling and charge injection to the SN. Therefore, the initial "0" value will always be significantly higher than ground for a PMOS MW. Similarly, the initial "1" value will always be significantly lower than V_{DD} for an NMOS write transistor. This limits the SN voltage range and degrades the overdrive of MR during readout, as well as the retention time. In a 2T GC, using the same device option for MR as for MW induces an additional voltage step in the same direction during read access, further impeding the performance. A hybrid cell, mixing NMOS and PMOS transistors [6, 11, 18, 23, 24], can be used to combat these effects, at a small area overhead for two different wells within each bitcell.

2.3.3.3 Peripheral Circuit Techniques

In addition to the choice of a GC topology and device options, several peripheral circuit techniques have been demonstrated to further improve system performance according to the target application. One simple and efficient technique is the employment of a sense buffer in place of a standard sense amplifier (SA) in low-power systems [13, 19, 24]. This implementation requires a larger RBL swing, trading off speed for area and process, voltage, temperature (PVT) sensitivity. The area trade-off is apparent in Fig. 2.4 as [19] shows exceptionally high array efficiency. Several other SA configurations have been demonstrated to deal with various design challenges. Chun et al. [6] overcome the problem of small RBL voltage swing by using a current mode SA featuring a cross-coupled PMOS latch and pseudo-PMOS diode pairs. Other SA designs include p-type gated diodes [20–22], offset compensating amplifiers [29], single-ended thyristors [28], and standard latches [31]. The most complex sensing scheme is used for multilevel GCs in [18, 23]: a successive approximation sensing scheme deciphers the four data levels.

Several publications [11, 13, 19, 24] discharge WBL during non-write operations to extend retention time that is worse for a stored "0" than a "1" with a PMOS WM. A "write echo refresh" (WER) technique was employed by Ichihashi et al. [11] to further reduce the WBL="1" disturbance. In this technique, the number of "1" write-back operations during refresh are counted and followed by WER operations (with oppositely biased WBL) to combat the disturbance. The authors of [5] recognized that the steady state level of a "1" and "0" is common (and much closer to "1" than to "0"), so they monitor this level and use it as the WBL voltage for writing a "1." This minimizes the "0" level disturbance without impeding the worst-case "1" level. For the system proposed in [6], WBL switching speed is the performance bottleneck, and therefore, a half-swing WBL is employed, improving the write speed and reducing the write power.

An issue that is rarely discussed in 2T GC implementations is the voltage saturation of RBL during readout. Depending on the implementation of MR, readout is achieved by either charging (NMOS) or discharging (PMOS) RBL. However, once RBL crosses a threshold (depending on the current ratio of the selected bitcell and the number of unselected cells), a steady state is reached. This phenomena not only limits the swing available for RBL sensing, but also causes static current dissipation that is present throughout the entire read operation. This is one of the phenomena which should be considered when choosing the appropriate V_{DD} for a low-power GC-eDRAM. Somasekhar et al. [31] combat this self clamping of RBL by explicitly clamping its voltage with designated devices.

2.3.4 Summary and Conclusions

This chapter reviewed and compared recently proposed GC-eDRAMs, categorizing them according to their target applications and overviewing the characteristics that make them appropriate for these applications. A closer look into the circuit design of these GC-eDRAMs provided further insight into the methods used to achieve the required design metrics through the use of different bitcell topologies, device options, technology nodes, and peripheral circuit implementations. To summarize briefly, the following best-practice guidelines should be used when designing GC-eDRAMs for future applications:

- High-V_T write access transistors for long retention times and low refresh power, in conjunction with area-efficient sense buffers for high array efficiency are most suitable to meet the storage requirements of ultra-low power (biomedical) VLSI SoCs.
- High-speed applications should use sensitive sense amplifiers to overcome small voltage differences, and should consider the use of low-V_T readout transistors for improved read access speed.
- Systems which frequently update their internal memories can sacrifice GC-eDRAM retention time for the benefits of high-speed access and high bandwidth.

References

1. Bushnell M, Agrawal V (2000) Essentials of electronic testing for digital, memory and mixed-signal VLSI circuits, chap 9.1. Springer, Berlin
2. Chang MT, Huang PT, Hwang W (2007) A 65nm low power 2T1D embedded DRAM with leakage current reduction. In: Proceedings of the IEEE SOC conference (SOCC), pp 207–210
3. Chun KC, Jain P, Lee JH, Kim C (2009) A sub-0.9V logic-compatible embedded DRAM with boosted 3T gain cell, regulated bit-line write scheme and PVT-tracking read reference bias. In: Proceedings of the IEEE symposium on VLSI circuits (VLSIC), pp 134–135
4. Chun KC, Jain P, Kim TH, Kim C (2010) A 1.1V, 667MHz random cycle, asymmetric 2T gain cell embedded DRAM with a 99.9 percentile retention time of 110 μ sec. In: Proceedings of the IEEE symposium on VLSI circuits (VLSIC), pp 191–192
5. Chun KC, Jain P, Lee JH, Kim C (2011) A 3T gain cell embedded DRAM utilizing preferential boosting for high density and low power on-die caches. IEEE J Solid State Circuits 46(6):1495–1505
6. Chun KC, Jain P, Kim TH, Kim C (2012a) A 667 MHz logic-compatible embedded DRAM featuring an asymmetric 2T gain cell for high speed on-die caches. IEEE J Solid State Circuits 47(2):547–559. doi:10.1109/JSSC.2011.2168729
7. Chun KC, Zhang W, Jain P, Kim C (2012b) A 2T1C embedded DRAM macro with no boosted supplies featuring a 7T SRAM based repair and a cell storage monitor. IEEE J Solid State Circuits 47(10):2517–2526
8. Giterman R, Teman A, Meinerzhagen P, Burg A, Fish A (2014) 4T gain-cell with internal-feedback for ultra-low retention power at scaled CMOS nodes. In: Proceedings of the IEEE international symposium on circuits and systems (ISCAS), pp 2177–2180
9. Giterman R, Teman A, Atias L, Fish A (2015) A soft error tolerant 4T gain-cell featuring a parity column for ultra-low power applications. In: Proceedings of the IEEE SOI-3D-subthreshold microelectronics technology unified conference (S3S), pp 1–2
10. Giterman R, Teman A, Meinerzhagen P, Atias L, Burg A, Fish A (2016) Single-supply 3T gain-cell for low-voltage low-power applications. IEEE Trans Very Large Scale Integr (VLSI) Syst 24(1):358–362
11. Ichihashi M, Toda H, Itoh Y, Ishibashi K (2005) 0.5 V asymmetric three-tr. cell (ATC) DRAM using 90 nm generic CMOS logic process. In: Proceedings of the IEEE symposium on VLSI circuits (VLSIC), pp 366–369

12. Ikeda N, Terano T, Moriya H, Emori T, Kobayashi T (2000) A novel logic compatible gain cell with two transistors and one capacitor. In: Proceedings of the IEEE symposium on VLSI technology (VLSIT), pp 168–169
13. Iqbal R, Meinerzhagen P, Burg A (2012) Two-port low-power gain-cell storage array: voltage scaling and retention time. In: Proceedings of the IEEE international symposium on circuits and systems (ISCAS), pp 2469–2472
14. Kaeslin H (2008) Digital integrated circuit design: from VLSI architectures to CMOS fabrication, 1st edn. Cambridge University Press, Cambridge
15. Kaku M, Iwai H, Nagai T, Wada M, Suzuki A, Takai T, Itoga N, Miyazaki T, Iwai T, Takenaka H, Hojo T, Miyano S, Otsuka N (2008) An 833MHz pseudo-two-port embedded DRAM for graphics applications. In: Proceedings of the IEEE international solid-state circuits conference (ISSCC), pp 276–613. doi:10.1109/ISSCC.2008.4523164
16. Kang S, Leblebici Y (2003) CMOS digital integrated circuits: analysis and design, 3rd edn. McGraw-Hill, New York
17. Karakonstantis G, Roth C, Benkeser C, Burg A (2012) On the exploitation of the inherent error resilience of wireless systems under unreliable silicon. In: Proceedings of the ACM/EDAC/IEEE design automation conference (DAC), pp 510–515
18. Khalid M, Meinerzhagen P, Burg A (2012) Replica bit-line technique for embedded multilevel gain-cell DRAM. In: Proceedings of the IEEE international NEWCAS conference, pp 77–80
19. Lee Y, Chen MT, Park J, Sylvester D, Blaauw D (2010) A 5.42 nW/kB retention power logic-compatible embedded DRAM with 2T dual-VT gain cell for low power sensing applications. In: Proceedings of the IEEE Asian solid state circuits conference (A-SSCC), pp 1–4
20. Luk W, Dennard R (2004) 2T1D memory cell with voltage gain. In: Proceedings of the IEEE symposium on VLSI circuits (VLSIC), pp 184–187. doi:10.1109/VLSIC.2004.1346552
21. Luk W, Dennard R (2005) A novel dynamic memory cell with internal voltage gain. IEEE J Solid State Circuits 40(4):884–894. doi:10.1109/JSSC.2004.842854
22. Luk W, Cai J, Dennard R, Immediato M, Kosonocky S (2006) A 3-transistor DRAM cell with gated diode for enhanced speed and retention time. In: Proceedings of the IEEE symposium on VLSI circuits (VLSIC), pp 184–185
23. Meinerzhagen PA, Andiç O, Treichler J, Burg AP (2011) Design and failure analysis of logic-compatible multilevel gain-cell-based DRAM for fault-tolerant VLSI systems. In: Proceedings of the IEEE/ACM great lakes symposium on VLSI (GLSVLSI), pp 343–346
24. Meinerzhagen P, Teman A, Mordakhay A, Burg A, Fish A (2012) A sub-VT 2T gain-cell memory for biomedical applications. In: Proc. IEEE subthreshold microelectronics conference (SubVT), pp 1–3. doi:10.1109/SubVT.2012.6404318
25. Meinerzhagen P, Teman A, Burg A, Fish A (2013a) Impact of body biasing on the retention time of gain-cell memories. IET J Eng 1:1–4
26. Meinerzhagen P, Teman A, Giterman R, Burg A, Fish A (2013b) Exploration of sub-VT and near-VT 2T gain-cell memories for ultra-low power applications under technology scaling. J Low Power Electron Appl 3(2):54–72. doi:10.3390/jlpea3020054
27. Meinerzhagen P, Bonetti A, Karakonstantis G, Roth C, Gürkaynak F, Burg A (2015) Refresh-free dynamic standard-cell based memories: application to a QC-LDPC decoder. In: Proceedings of the IEEE international symposium on circuits and systems (ISCAS), pp 1426–1429
28. Park YS, Blaauw D, Sylvester D, Zhang Z (2012) A 1.6-mm^2 38-mW 1.5-Gb/s LDPC decoder enabled by refresh-free embedded DRAM. In: Proceedings of the IEEE symposium on VLSI circuits (VLSIC), pp 114–115
29. Somasekhar D, Lu SL, Bloechel B, Dermer G, Lai K, Borkar S, De V (2005) A 10 Mbit, 15 GBytes/sec bandwidth 1T DRAM chip with planar MOS storage capacitor in an unmodified 150 nm logic process for high-density on-chip memory applications. In: Proceedings of the IEEE European solid-state circuits conference (ESSCIRC), pp 355–358

30. Somasekhar D, Ye Y, Aseron P, Lu SL, Khellah M, Howard J, Ruhl G, Karnik T, Borkar S, De V, Keshavarzi A (2008) 2 GHz 2 Mb 2T gain-cell memory macro with 128 GB/s bandwidth in a 65 nm logic process. In: Proceedings of the IEEE international solid-state circuits conference (ISSCC), pp 274–613
31. Somasekhar D, Ye Y, Aseron P, Lu SL, Khellah M, Howard J, Ruhl G, Karnik T, Borkar S, De V, Keshavarzi A (2009) 2 GHz 2 Mb 2T gain cell memory macro with 128 GBytes/sec bandwidth in a 65 nm logic process technology. IEEE J Solid State Circuits 44(1):174–185
32. Zhang W, Chun KC, Kim CH (2010) Variation aware performance analysis of gain cell embedded DRAMs. In: Proceedings of the ACM/IEEE international symposium on low power electronics and design (ISLPED), pp 19–24

Chapter 3
Retention Time Modeling: The Key to Low-Power GC-eDRAMs

3.1 Introduction

Gain-Cell eDRAM (GC-eDRAM) is an interesting, high-density alternative to SRAM and conventional 1T-1C eDRAM for a large range of VLSI system-on-chip (SoC) applications, including ultra-low power systems such as biomedical implants [17], wireless communications systems (e.g., LDPC decoders [19]), and high-performance microprocessors [3, 7], as discussed in detail in Chap. 2. GC-eDRAM has many assets, including small bitcell size, low bitcell leakage current, compatibility with standard digital CMOS processes (often referred to as "logic compatibility"), and inherent suitability for two-port memory implementations [22, 26]. However, GC-eDRAM is also characterized by a primary drawback—the need for power-hungry refresh cycles. Several previous studies based on silicon measurements [7, 16] unveiled large spreads of the per-cell retention time (RT) across a GC array. Unless spare columns, spare rows, or error correction codes (ECC) [25] are used, the worst GC dictates the refresh rate of the entire array [5]. Longer RTs and a sharper per-cell RT distribution are desirable for less frequent refresh and thus lower power consumption [24].

While Monte Carlo (MC) circuit simulations and silicon measurements can be used to obtain per-cell RT maps [10, 16, 23], these methods, unfortunately, do not unveil the factors that limit the RT and the circuit parameters, which are responsible for the large RT spreads. Furthermore, extensive MC circuit simulations and silicon measurements do not allow for early design considerations, such as trading off read failure probability for refresh power, either for fault-tolerant applications or in conjunction with spare columns/rows or ECC for error-free applications [24]. In [8], the influence of different design parameters on the RT of GC-eDRAM was examined; however, this work did not provide any analytical model to gain deeper insights, nor did not relate to the statistical distribution of RT. A model of retention time was presented in [9], but it related only to the conventional 1T-1C eDRAM bitcell and not to GC-eDRAM. In addition, the model of [9] referred only

P. Meinerzhagen et al., *Gain-Cell Embedded DRAMs for Low-Power VLSI Systems-on-Chip*, DOI 10.1007/978-3-319-60402-2_3

to the nominal value and not to the statistical distribution of retention time. In [24], the potential of exploiting the RT distribution of GC-eDRAM for fault-tolerant applications was explored, but this was solely based on MC simulations and silicon measurements. Briefly, in the open literature, there is a lack of simple analytical models for GC-eDRAM retention time and its statistical distribution across the array. Furthermore, there are no previous studies explaining the large GC-eDRAM retention time spreads, taking into account the variability of all circuit parameters.

This chapter derives an analytical model for the statistical RT distribution of GC-eDRAM to provide a tool for the design of GC-eDRAM bitcells and to understand the underlying trade-offs in the design parameters. The model accounts for parametric variations in devices (write and read transistors) and the storage node capacitor. The accuracy of the analytical model is verified by means of statistical MC circuit simulation, worst-case distance point analysis, and through silicon measurements of an 0.18 µm GC-eDRAM test chip. Furthermore, the model is validated in an advanced 28 nm node (see Chap. 6), showing that the proposed model holds, in light of technology scaling. In addition, a detailed sensitivity analysis is presented, identifying the circuit parameters with the largest impact on the per-cell RT spread. The analytical model for RT distribution can be used for the design of various types of VLSI SoCs, including low-power VLSI SoCs, in the following ways:

1. For low-power VLSI SoCS, including biomedical sensor nodes where correct circuit operation is of utmost importance, the model is a convenient means to set a safety guardband on the refresh rate according to the memory array size and manufacturing yield requirements.
2. For wireless communications systems [11] or other systems that are inherently resilient to a small number of hardware defects [28] (e.g., video and image processing), the retention time model allows designers to study the trade-off between read failure probability and refresh power by means of varying the refresh rate [18, 24].
3. For all VLSI SoC applications (error-free and error-resilient), especially if large storage capacities are required, it is important to identify the main cause for the large RT spread in prior-art GC-eDRAM implementations in order to eventually narrow this distribution and achieve longer data retention intervals and consequently lower refresh power [3].

3.2 Choice of Basic 2T GC-eDRAM Bitcell

As expatiated on in Chap. 2, several topologies of GC-eDRAM have been proposed, consisting of 2–4 transistors and in some cases an additional MOSCAP or diode to increase the storage capacitance [10, 22]. This chapter focuses on the 2T GC, shown in Fig. 3.1, since it comprises the basic structure that is common to all topologies: a write transistor (MW) and a storage/read transistor (MR). In addition, the 2T

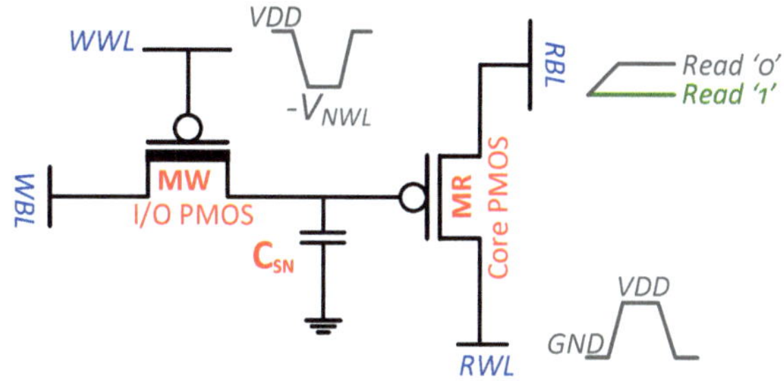

Fig. 3.1 Schematic representation of the basic 2T gain-cell used to develop the retention time model. Signal waveforms for read and write access are shown as well

bitcell has the smallest footprint among all GC-eDRAM bitcells, making it the most interesting bitcell for the implementation of high-density memories [21].

The write and read transistors of the 2T GC can be implemented with either PMOS or NMOS devices, trading off leakage, access speed, and area requirements, as discussed in detail in Chap. 6. In this chapter, we consider an all-PMOS version, implemented with a high-V_T I/O PMOS device as MW in order to reduce the RT limiting leakage currents, and a standard-V_T PMOS device as MR in order to eliminate area requirements of a within-cell well separation.

In this topology, the write operation is performed by applying a negative under-drive voltage ($-V_{NWL}$) to the write word-line (WWL) to overcome the threshold voltage (V_T) drop when discharging the storage node (SN) during a write "0" operation. For read operations, the read bit-line (RBL) is pre-discharged and the read word-line (RWL) is pulsed to the positive supply (V_{DD}). On the one hand, when the bitcell stores data "1," MR will be in cutoff and RBL will remain discharged. On the other hand, if data "0" is stored, MR will turn on and charge RBL past a read threshold. For clarity, the control signals are illustrated in Fig.3.1 next to the 2T GC schematic drawing. It is important to note that for the chosen 2T GC topology, data "0" deteriorates much faster than data "1" [15], and is therefore considered throughout this chapter as the limiting bitcell state for calculating RT. The refresh rate of the memory is determined by the worst-case retention time, which occurs when WBL is held at the opposite level of the stored data. This biases MW with a maximum V_{DS}, leading to maximum sub-V_T leakage, due to drain-induced barrier lowering (DIBL). While the continuous biasing of WBL at such a worst-case level is highly unlikely, especially when methods for reducing this situation are applied [23], it still needs to be considered as the RT limiter.

Note that a positively boosted gate voltage can be applied to WWL to reduce the sub-V_T leakage through MW [27]; however, this is accompanied by additional issues, such as the need to provide a third biasing voltage, gate leakage and stress on MW, and gate induced drain leakage (GIDL), which further complicates the circuit design and the RT model. Therefore, throughout the remainder of this chapter, we assume that WWL is held at V_{DD} during standby and read cycles.

3.3 Analytical GC-eDRAM Retention Time Model

3.3.1 *Definition of Retention Time*

While it is clearly essential to initiate a refresh operation on a GC-eDRAM array prior to data loss, the accurate definition of RT and its methods of simulation vary across different publications [22]. To avoid any confusion, the definition of data retention time (DRT) used throughout this chapter is provided hereafter, followed by a new metric, effective data retention time (EDRT), which is introduced to both take into account the performance requirements of the memory array, as well as to assist in the development and proof of the analytical model.

3.3.1.1 Data Retention Time (DRT)

For the purpose of the herein presented model, we define DRT as the time interval after a write operation at which the voltage difference between the stored data "1" (D1) and stored data "0" (D0) reaches a lower limit for a sufficient sensing margin. In the considered 0.18 μm CMOS technology with $V_{DD} = 0.8$ V, this limit is chosen to be $\delta V_{DRT} = V_{D1} - V_{D0} = 0.5$ V. Note that this value can be adapted according to the technology node, application, and sensitivity of the sensing circuitry.

3.3.1.2 Effective Data Retention Time (EDRT)

While the DRT metric defines the maximum retention time before losing the stored data value, it does not take into account the operating frequency (read speed) required by the application. Therefore, a new metric, EDRT, is introduced, defined as the time interval following a write operation still enabling a correct readout for a given read access time requirement. The read access time is determined by the time it takes a cell holding a "0" to charge the RBL to a level that can be detected by the sensing circuitry. Even though this is highly dependent on the RBL capacitance (number of rows in the array) and the chosen sensing circuitry, the current drive of MR that is required to provide this access time can be extracted from a given architecture. We use the overdrive of MR (corresponding to the deteriorated level of D0, or equivalently D1 for an NMOS MW) that provides this read current as the threshold voltage V_{EDRT} to measure the EDRT of a given bitcell. In the current analysis, we extracted the threshold from the test array to be $V_{EDRT} = V_{D0} = 0.270$ V to ensure compliance with an operating frequency of 1 MHz. However, this value should be adapted for any other CMOS technology, array architecture, and operating frequency.

While DRT is more commonly used in the literature, EDRT is easier to model analytically since it considers only the charging of SN, as opposed to DRT, which considers both SN charging and discharging. However, EDRT will be shown to closely track DRT, and therefore, it is an appropriate alternative to DRT for circuit design considerations.

3.3.2 *Analytical Model of Nominal EDRT*

This section derives an analytical model of GC-eDRAM EDRT, taking into account all circuit parameters of the all-PMOS 2T GC. While these circuit parameters follow a statistical distribution, the current section assumes nominal values for all parameters to derive the nominal value of EDRT.

The proposed EDRT model calculates the time t = EDRT it takes for the voltage on the storage capacitor V_{SN} to charge beyond the failure voltage V_{fail}. This time period is derived from the fundamental integral relationship between voltage and current of the storage capacitor (C_{SN}):

$$V_{\mathrm{SN}}(\mathrm{EDRT}) = V_{\mathrm{fail}} = \frac{1}{C_{\mathrm{SN}}} \int_{t_0}^{\mathrm{EDRT}} i_{\mathrm{SN}}(\tau)d\tau + V_{\mathrm{SN}}(t_0), \tag{3.1}$$

where i_{SN} is the leakage current (dominated, in the present 0.18 μm CMOS case, by sub-V_{T} leakage, I_{sub}), which charges the storage capacitor; and $V_{\mathrm{SN}}(t_0)$ is the voltage on C_{SN} at the initial time t_0.

The initial value on the SN following a write $V_{\mathrm{SN}}(t_0)$ is highly dependent on the WWL voltage boost (V_{NWL}), the write pulse transition and duration, and the coupling capacitance between WWL and SN ($C_{\mathrm{WWL\_SN}}$), as elaborated upon extensively in [17].

To extract the worst-case EDRT, the presented model assumes that immediately following a write "0" operation ("1" for an NMOS MW), MW is cut-off with WWL = V_{DD}, and WBL is biased to V_{DD} (0 for an NMOS MW) to cause a worst-case deterioration of the D0 level (D1 level for an NMOS MW), primarily due to sub-V_{T} leakage through MW [20]. The sub-V_{T} leakage is modeled according to the well-known EKV sub-V_{T} current equation [4]:

$$I_{\mathrm{sub}} = \frac{W}{L} I_{s0} \left(1 - e^{\frac{-V_{\mathrm{DS}}}{v_t}}\right) e^{\frac{V_{\mathrm{GS}}-V_{\mathrm{th}}-V_{\mathrm{off}}}{n \cdot v_t}}, \tag{3.2}$$

where W and L are the device width and length, respectively; V_{DS} and V_{GS} are the drain-to-source and gate-to-source voltages, respectively; v_t is the thermal voltage; V_{th} is the threshold voltage; V_{off} is the offset voltage that determines the drain current at $V_{\mathrm{GS}} = 0$; and n is the subthreshold swing coefficient. Furthermore, I_{s_0} is defined as:

$$I_{s_0} = \mu_0 \sqrt{\frac{q\varepsilon_{\mathrm{si}} N_{\mathrm{ch}}}{2\phi_s}} v_t, \tag{3.3}$$

Table 3.1 MW parameters used for EDRT calculation

Parameter	Value	Parameter	Value
$V_{fail} = V_{SN}(\text{EDRT})$	0.27 V	$V_{SN}(t_0)$	0 V
W	240 nm	L	340 nm
V_{SG}	0 V	n	1.51
C_{SN}	3 fF	$V_{th,p}$	−720 mV
I'_{s0}	260 nA	v_t	26 mV

where q is the electric charge; μ_0 is the mobility at nominal temperature; N_{ch} is the channel doping concentration; and ϕ_s is the surface potential.

During retention periods, we assume that $V_{DS} \gg v_t \simeq 26\,\text{mV}$, and consequently that the sub-V_T leakage is constant.[1]

Substituting (3.2) in (3.1) yields the following analytical model of EDRT:

$$\text{EDRT} = \frac{[V_{SN}(\text{EDRT}) - V_{SN}(t_0)]\, C_{SN} L}{W I'_{s0} \exp\left(\frac{V_{GS}}{n v_t}\right)} \exp\left(\frac{V_{th}(L)}{n v_t}\right) \tag{3.4}$$

with:

$$I'_{s0} = I_{s0} e^{-\frac{V_{off}}{n v_t}}. \tag{3.5}$$

All parameters used in (3.4) and (3.5) refer to MW. Also, all parameters used in this chapter for EDRT calculation are summarized in Table 3.1.

3.3.3 *Statistical Distribution of EDRT*

Based on the analytical model for nominal EDRT (3.4), this section derives the statistical distribution of EDRT, taking into account the statistical distributions of primary circuit parameters. From the model presented in (3.4) and (3.5), it can be seen that the EDRT has an exponential dependence on the V_T of MW (V_T^{MW}). Moreover, a simulation-based sensitivity analysis unveils that V_T^{MW} is responsible for more than 50% of the EDRT variation (see Sect. 3.6 for details). From these two observations, and from the fact that V_T follows a Gaussian distribution arising from process parameter variations, it can be concluded that a log-normal distribution is an appropriate description for the statistical distribution of EDRT.

In fact, given a normally distributed random variable X with a mean of μ and standard deviation σ ($X \sim \text{N}(\mu, \sigma^2)$) and a function $Y = a \cdot e^{bX}$ (where a and b are constants), then function Y is log-normally distributed with coefficients (referred to as location parameter and scale parameter [1]) given by:

$$Y \sim \log N(b\mu + \ln(a), (b\sigma)^2). \tag{3.6}$$

[1] This holds true for the applicable range of $V_{SN}(t_0) < V_{SN} < V_{EDRT}$.

Therefore, based on (3.4), the coefficients of the log-normal EDRT distribution can be expressed as follows:

$$\begin{aligned} \text{EDRT} \sim \Bigg[\log N \Bigg(& \frac{1}{nv_t}\mu(V_{\text{th}}) \\ & + \ln\left(\frac{[V_{\text{SN}}(\text{EDRT}) - V_{\text{SN}}(t_0)]C_{\text{SN}}L}{WI'_{s0}\exp\left(\frac{V_{\text{SG}}}{nv_t}\right)}\right), \\ & \left(\frac{\sigma(V_{\text{th}})}{nv_t}\right)^2\Bigg)\Bigg]. \end{aligned} \tag{3.7}$$

A numeric evaluation of (3.7), extracting the mean value and the variance of V_{T}^{MW} from an 0.18 µm CMOS design kit, accounting for global and local variations, leads to:

$$\text{EDRT [s]} \sim \log N(-0.95, 0.36) \tag{3.8}$$

3.4 Model Validation Through Circuit Simulations

This section verifies the accuracy of the analytical models for nominal EDRT and its statistical distribution by comparison with MC circuit simulation results and worst-case distance analysis in 0.18 µm CMOS technology.

3.4.1 Nominal EDRT

The analytical model for the nominal EDRT, presented in (3.4), is verified by means of transistor-level circuit simulation of the 2T GC shown in Fig. 3.1. The circuit is implemented in a 0.18 µm CMOS technology node, using a high-V_{T} I/O PMOS device as MW and a standard-V_{T} PMOS device as MR. The simulations are based on nominal process parameter values, with the exception of V_{T}^{MW}, which is swept to represent $\pm 3\sigma$ global and local process variations. Simulations are carried out for a typical-typical (TT) process corner (used also as a starting point for V_{T}^{MW}, prior to applying parametric variations), a supply voltage of 0.8 V and a temperature of 27 °C.

Figure 3.2 shows the simulated EDRT ("×" markers) as a function of V_{T}^{MW}, as compared to the analytical EDRT model, represented by the solid blue line. The model coefficients were computed based on (3.4), using the primary circuit and physical parameters of Table 3.1; however, the same coefficients can also be found

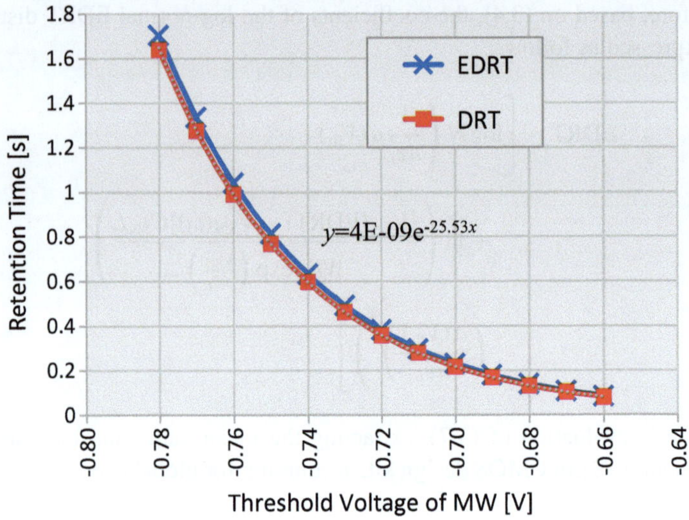

Fig. 3.2 Modeled and simulated retention time (EDRT and DRT) versus V_T^{MW}

by fitting an exponential curve to the simulated data points. Clearly, the model very closely matches the simulated data, with a maximum modeling error of less than 6%. This plot confirms the exponential dependence of EDRT on V_T^{MW}, as shown in (3.4), which results in a strong impact of V_T^{MW} variability on the spread of the EDRT.

Both above-mentioned methods to determine the model coefficients lead to the same coefficient values, within a reasonable numeric precision. Simulating as few as 10 data points, followed by exponential curve fitting, might often be a more convenient and straightforward approach to determine the coefficients of the EDRT model than deriving the same coefficients from primary circuit and physical parameters.

Figure 3.2 also shows the simulated DRT ("□" markers), which closely follows the EDRT. The solid red line is an exponential fit to the simulated DRT data points, and suggests that DRT also has an exponential dependence on V_T^{MW}, similar to EDRT. While this chapter did not explicitly and analytically model DRT, an exponential DRT model can be easily obtained through curve fitting.

Figure 3.3 shows the relationship between EDRT and the metal stack capacitor C_{SN}, as well as W and L of MW (W_{MW} and L_{MW}, respectively). These three parameters are varied by 15% around their nominal values (only one parameter is shifted at a time), as shown on the x-axis. The data points are extracted from circuit simulation, while the solid lines correspond to the analytical EDRT model. Both the analytical EDRT model and the simulations lead to the same conclusions:

1. EDRT is proportional to C_{SN} and inversely proportional to W_{MW}. This is true as long as the influence of W_{MW} on C_{SN} is negligible. Scaling C_{SN} and W_{MW} by the same factor leaves the retention time unchanged.

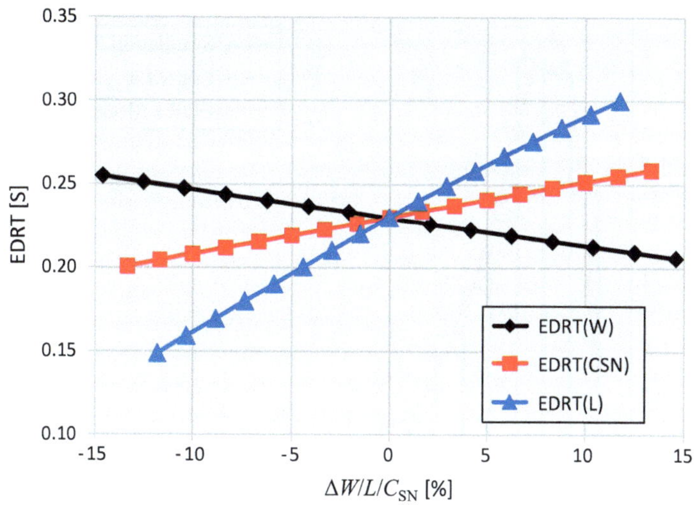

Fig. 3.3 EDRT as a function of the metal stack capacitor and the width and length of MW

2. EDRT increases proportionally with L_{MW}. In addition, L_{MW} has a stronger impact on EDRT than W_{MW} and C_{SN}. A detailed sensitivity analysis of these parameters is provided in Sect. 3.6.

In summary, the analytical EDRT model of (3.4) firmly tracks the simulated data over sweeps of V_T^{MW}, W_{MW}, and C_{SN}, with a maximum modeling error of 6%.

3.4.2 Statistical EDRT Distribution

After having confirmed the accuracy of the nominal EDRT model, this section compares the statistical EDRT distribution model, given in (3.6) and (3.7), with MC circuit simulations in 0.18 µm CMOS technology. Recall that the random variable EDRT was found to follow a log-normal distribution: EDRT $\sim \log N(-0.95, 0.36)$. The MC analysis is based on statistical process parameter distributions, provided by the 0.18 µm foundry. Both global and local parametric variations are accounted for by taking 10,000 EDRT samples.

Figure 3.4 shows the statistical EDRT distribution obtained from MC simulation (blue bars), as well as the analytically modeled log-normal EDRT distribution (red solid line). In addition, the dashed blue line shows a simple log-normal fit to the simulated distribution. Clearly, the analytical EDRT distribution model corresponds well with the simulated and fitted data, with modeling errors of less than 4% for the mean and standard deviation of the EDRT (μ_{EDRT} and σ_{EDRT}, respectively).

The proposed EDRT model is intended to be a design model that compromises accuracy for simplicity to enable the bitcell designer to comprehend the inherent

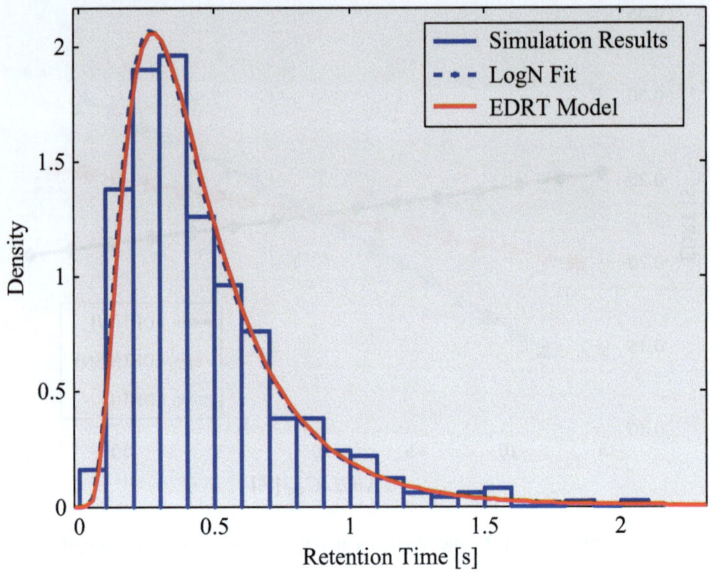

Fig. 3.4 Statistical EDRT distributions: model and MC simulations

trade-offs by modifying a given circuit parameter. The main accuracy compromise occurs at the far end of the EDRT distribution, where a more complex model is required to accurately capture the extreme values. However, to evaluate the usefulness of the model to estimate extreme values, the tail of the model was compared to extreme value extraction, using the Cadence worst-case distance (WCD) high-yield estimation tool. The EDRT model showed less than a 10% deviation from the simulated values, demonstrating its usefulness, even for estimation of retention time in extreme cases.

3.5 Model Validation Through Silicon Measurements of 0.18 μm CMOS Test Arrays

This section verifies the high accuracy of the analytical EDRT model and the MC simulations by comparison with silicon measurements of a test chip fabricated in 0.18 μm CMOS technology.

3.5.1 Test Chip Design

A 64 × 32 (2 kb) GC-eDRAM macro using the GC topology, shown in Fig. 3.1, was fabricated in a 0.18 μm CMOS technology. The memory array was divided into four sub-arrays of 16×32 bits (referred to as: *Nominal, NoCap, SmallCap,* and *LargeMW*), in order to study the influence of different circuit parameters on retention time and its distribution. This allows the verification of the analytical model and the MC circuit simulations for a range of GC configurations. The exact parameters of the four 2T GC configurations are explained in the following:

1. *Nominal*: This is the baseline GC with a minimum-size I/O MW, minimum-size core MR, and a 3 fF C_{SN} (according to layout extraction) built from the available metal stack. The nominal GC sub-array is used to confirm the random, as opposed to systematic, per-cell retention time distribution, which, according to simulations and modeling, is mostly influenced by V_T^{MW} variability (see Sect. 3.6).
2. *NoCap*: The second GC configuration has the same MW and MR as the nominal GC, but has no metal-stack capacitor. Therefore, C_{SN} consists only of the inherent MOSCAP and diffusion capacitance. This GC is targeted at isolating the variability of V_T^{MW} from that of C_{SN} in order to separately evaluate the impact of these two parameters.
3. *SmallCap*: The third GC configuration is identical to the nominal GC, albeit with a reduced metal stack, providing only 2 fF of storage capacitance. This sub-array is used to further study the impact of C_{SN}, the main circuit parameter for conventional eDRAM, on RT and its distribution.
4. *LargeMW*: This last GC configuration has an up-sized MW compared to the nominal cell, with a 25% increase in both W_{MW} and L_{MW}. This GC uses the nominal 3 fF C_{SN}. The goal of this GC is to study the impact of decreased V_T^{MW} variance on the per-cell RT distribution.

Figure 3.5 shows the layout of the *Nominal* GC, as well as the layout of the entire GC-eDRAM test array. The four sub-arrays are clearly visible and annotated. The test array macrocell is implemented as a two-port memory with separate read and write ports. The read and write address decoders are implemented as two-stage decoders, containing a 3:8 pre-decoder followed by a decoder using 2-input NOR gates. The read path contains a dynamic sense inverter. To facilitate testing, a frame of flip-flops around the macrocell holds addresses and data constant for a full clock cycle. Furthermore, the test chip contains a built-in self-test (BIST), a scan chain interface to access the full test array, and direct access to a sub-array.

3.5.2 Measurement Results

In order to perform silicon measurements, five packaged test chips were mounted onto a PCB via a burn-in test socket. An Agilent 16822A logic analyzer with built-in pattern generator was used to extract retention time information, by interfacing with and triggering the BIST. The array was operated at a supply voltage of 0.8 V, while the read and write ports were clocked at 1 MHz.

Figure 3.6 shows the measured RT (for a read frequency of 1 MHz) as well as the simulated EDRT and DRT (also for a read frequency of 1 MHz). Measured data is taken from a single typical (TT) die, while the MC simulations correspond to local (within-die) parametric variations in a TT corner. The figure further shows

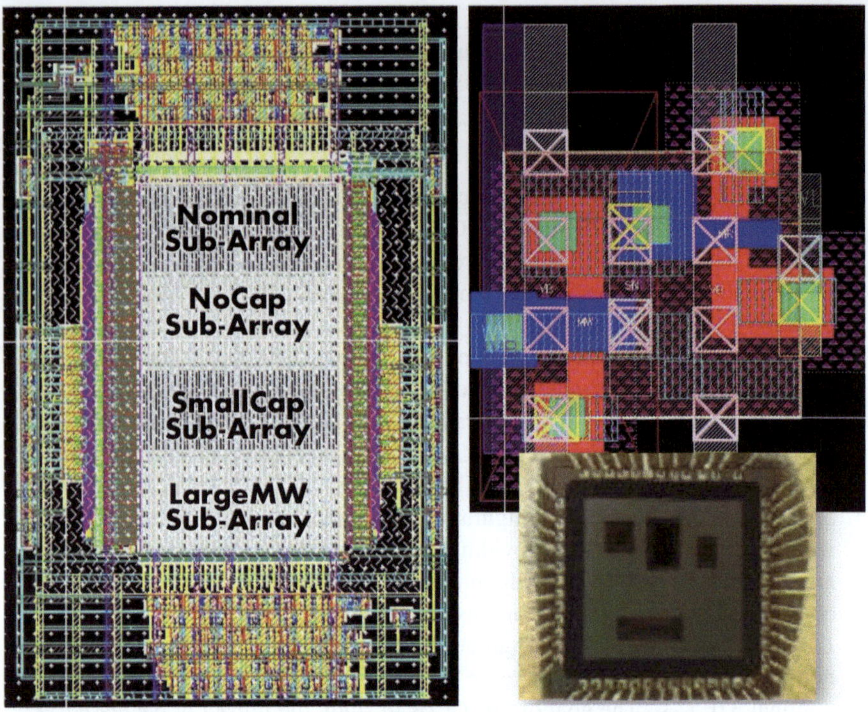

Fig. 3.5 Complete test array macrocell with sub-arrays (*left*), layout of nominal GC (*top left*), and chip photograph (*bottom left*)

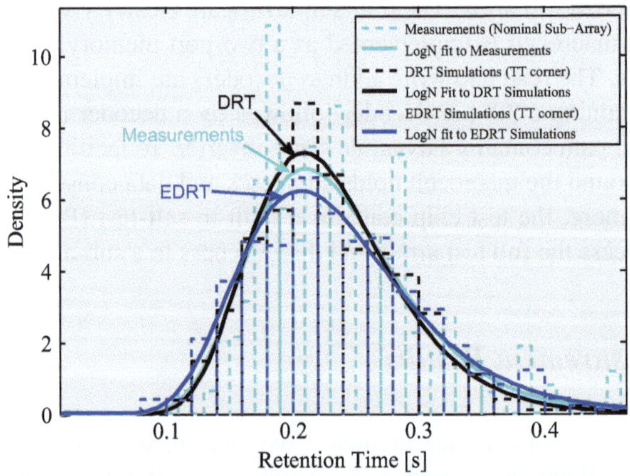

Fig. 3.6 Silicon measurements of retention time compared to simulations and analytical modeling

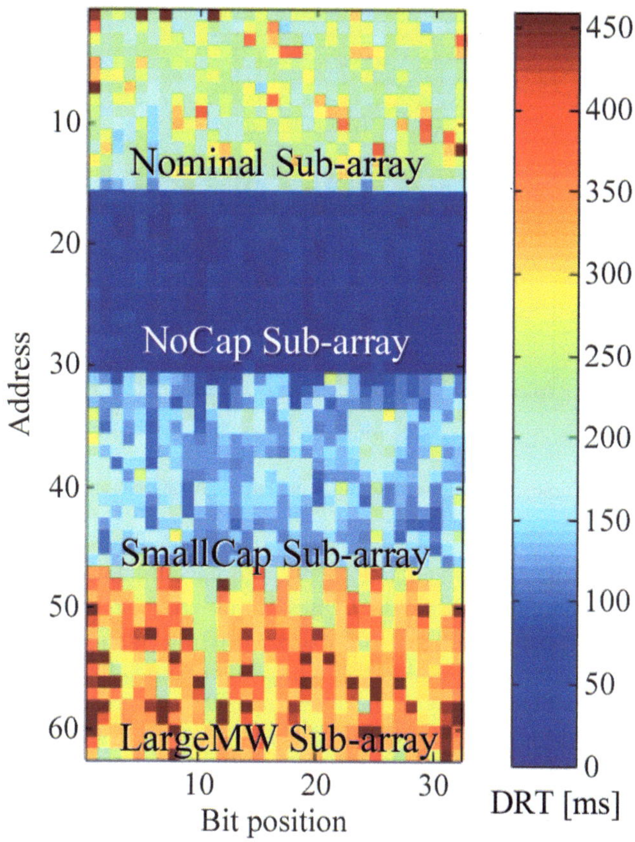

Fig. 3.7 Data retention time map of four measured GC sub-arrays, including mean and variance of log-normal distribution of each sub-array

the log-normal fit to the measured and simulated data, with the fitted EDRT curve corresponding to the analytical model of (3.7). It is apparent from Fig. 3.6 that the measured retention time follows a log-normal distribution and that there is a good match between silicon measurements, MC simulations, and modeling results.

The measured retention time distribution lies between the simulated EDRT and DRT distributions. In fact, the EDRT and DRT definitions become almost equivalent for the slow considered read frequency (1 MHz), and coincide with the silicon measurements equally well. The silicon measurements confirm the possibility of precisely predicting the RT spread using a log-normal distribution, with coefficients extracted from a few simulated or measured data points or derived from the analytical model in Eq. (3.7).

Figure 3.7 shows the measured per-cell RT for the entire 64×32 array. The RT of each GC is shown in a color code at the physical location of each GC. The four sub-arrays using different GC configurations are immediately seen to exhibit different retention timcs.

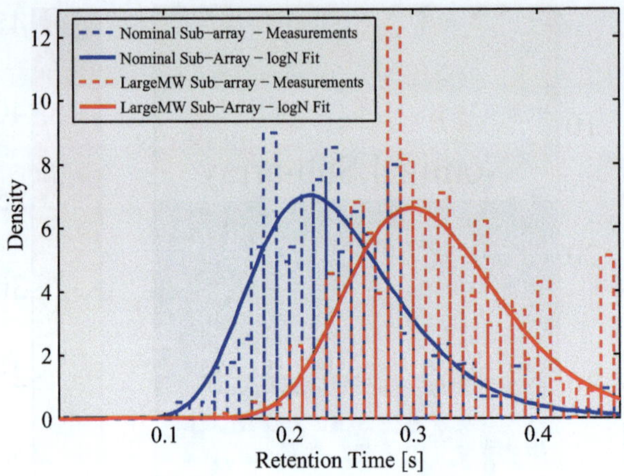

Fig. 3.8 Retention time distributions of nominal and largeMW GC sub-arrays

For example, the *NoCap* sub-array appears to have the lowest retention times, centered at roughly 24 ms, while the *LargeMW* sub-array has the longest retention times with an average value of around 320 ms. In all four sub-arrays, the per-cell RT clearly appears to be distributed randomly; in other words, there is no systematic RT distribution pattern for any of the sub-arrays. This observation strengthens the assumption that the RT spread is mostly caused by variation of V_T^{MW}, which in turn is mostly caused by random dopant fluctuations (RDF).

Figure 3.8 shows the measured RT distributions of the *Nominal* and *LargeMW* GC sub-arrays, and their log-normal fits. Due to the larger transistor dimensions in the *LargeMW* GC sub-array, V_T^{MW} varies less, and therefore the σ_{RT} parameter of the log-normal distribution is only 0.198, as compared to 0.254 for the *Nominal* GC sub-array. Furthermore, the μ_{RT} parameter of the distribution also improves from -1.46 for the *Nominal* GC sub-array to -1.16 for the *LargeMW* GC sub-array due to the higher V_T^{MW} caused by increased L_{MW} (short channel effect). As the mean and variance of the log-normal distribution are a function of both μ_{RT} and σ_{RT}, as shown in (3.9), there is a trade-off between the mean and the variance of the log-normal distribution.

$$\begin{aligned} \text{Mean} &= e^{\mu+\frac{\sigma^2}{2}} \\ \text{Variance} &= (e^{\sigma^2}-1)e^{2\mu+\sigma^2} \end{aligned} \tag{3.9}$$

As μ_{RT} increases, e.g., due to a larger C_{SN} or longer L_{MW}, the RT variance also increases. Therefore, in Fig. 3.8, it can be seen that the mean RT improved from 0.238 s for the *Nominal* GC sub-array to 0.316 s for the *LargeMW* GC sub-array, but the RT variance, which originally was estimated to be smaller due to the larger dimensions of the write transistor, stays almost the same, due to the influence of μ_{RT} (0.0038 s in the *Nominal* compared to 0.0040 s in the *LargeMW* GC sub-array). Furthermore, if C_{SN} is smaller, the mean and variance both decrease, and as a result, a sharper distribution is achieved (as observed in the measurements of the *SmallCap* and *NoCap* GC sub-arrays), at the cost of lower retention time. In conclusion, GC design needs to consider both μ_{RT} and σ_{RT} in order to find the best trade-off between the mean and the variance of the retention distribution. According to our mathematical derivations, increasing V_T^{MW} increases the mean RT, but unfortunately it increases the RT variance as well.

In summary, the measured retention time distribution is well approximated by the EDRT model and the MC simulations, and the measured per-cell retention time is randomly distributed. Up-scaling MW improves retention time and sharpens its distribution.

3.6 Sensitivity Analysis of GC-eDRAM Retention Time

The previous section, as well as previous publications that reported the per-cell retention time of fabricated GC arrays [6, 16], show very large retention time spreads. Unfortunately, unless redundancy or error correction are integrated into or built around the array, the worst cell in terms of DRT ultimately dictates the global refresh rate. Therefore, it is preferable to have a sharp per-cell RT distribution and move the worst cell close to a nominal cell. In order to do this, it is crucial to understand the influence of specific circuit parameters on the RT spread. Accordingly, a sensitivity analysis of basic design parameters on RT is presented in this section.

3.6.1 Plackett-Burman Design of Experiment (PB-DOE)

The RT sensitivity analysis presented below was performed using the Plackett-Burman Design of Experiment (PB-DOE) [14]. This is a powerful statistical tool to estimate the influence of individual process parameters in multivariate systems. The PB-DOE method has been verified as an effective method to identify the most sensitive process parameters that cause variability in device performance [13]. This method is based on the concept of orthogonal arrays and needs to run N simulations to analyze $N - 1$ parameters [14].

Reference [12] provides guidelines for constructing a PB-DOE matrix for implementing an N-parameter sensitivity analysis. In the example shown in Table 3.2, these guidelines were used to construct a seven parameter matrix, with P1 through P7 representing the examined process parameters, and "+" and "−" representing the maximum and minimum values of each process parameter, respectively. Each row in the table determines the value of each one of the seven parameters for each simulation run. Y1 through Y8 indicate the circuit response for each specific set of process parameters. The sensitivity of circuit performance to each of these parameters is the ratio between the sensitivity of the n-th process parameter for a

Table 3.2 Example of PB-DOE matrix for seven factors

Run	P1	P2	P3	P4	P5	P6	P7	Y
1	+	+	+	−	+	−	−	Y1
2	−	+	+	+	−	+	−	Y2
3	−	−	+	+	+	−	+	Y3
4	+	−	−	+	+	+	−	Y4
5	−	+	−	−	+	+	+	Y5
6	+	−	+	−	−	+	+	Y6
7	+	+	−	+	−	−	+	Y7
8	−	−	−	−	−	−	−	Y8

given response ($\mathrm{SS(P}_n\mathrm{)}$), to the total sum of all $\mathrm{SS(P}_n\mathrm{)}$ due to all process parameters ($\mathrm{SS_t}$). This can be written as:

$$\mathrm{SS(P}_n\mathrm{)} = [\mathrm{avg(Y+)} - \mathrm{avg(Y-)}]^2 \tag{3.10}$$

$$\mathrm{SS_t} = \mathrm{SS(P_1)} + \mathrm{SS(P_2)} + \cdots + \mathrm{SS(P}_n\mathrm{)}, \tag{3.11}$$

where avg(Y+) is the average value of all the simulation runs in which the corresponding process parameter was high (+), and avg(Y−) is the average value of all the simulation runs in which the corresponding process parameter was low (−).

3.6.2 PB-DOE Applied to GC-eDRAM RT

Several parameters that are susceptible to process variations contribute to the wide spread of RT. These parameters can be divided into three primary categories, according to the circuit metric they affect:

1. sub-V_T leakage (I_{sub}) through MW: affected by V_T, gate oxide thickness (t_{ox}), W, and L of MW.
2. Drive strength of MR: affected by V_T, t_{ox}, W, and L of MR.
3. Magnitude of C_{SN}: diffusion capacitance (C_{GD}) of MW, gate capacitance (C_{GC}), t_{ox}, W, and L of MR, and interconnect capacitance, such as that achieved through implementing a metal stack above the SN.

To apply the PB-DOE method of sensitivity analysis of 11 parameters to GC-eDRAM RT, the following 10 transistor-related parameters were analyzed: L, W, V_T, t_{ox}, and gate-drain overlap capacitance coefficient (C_{gdo}) for both MW and MR. The high and low values of each parameter were determined according to the 3σ value accounting for global and local process parameter variations. In order to capture the sensitivity of retention time to C_{SN} variability due to variations in the fabrication of the metal stack, this parameter was added to the analysis as the 11th parameter. Since detailed capacitance variability reports were not available from the foundry, a 10% deviation from the nominal value was assumed.

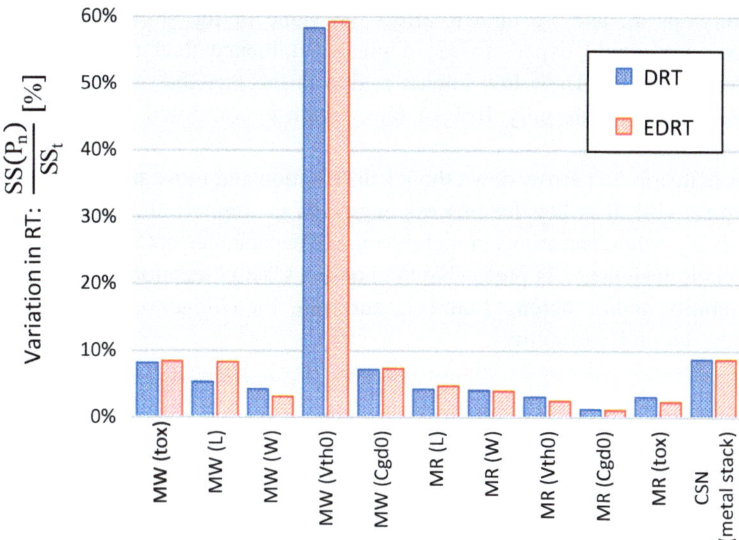

Fig. 3.9 Sensitivity analysis of DRT and EDRT on circuit parameters: the y-axis shows $SS(P_n)/SS_t$

The sensitivity analysis results are presented in Fig. 3.9, clearly showing that the dominant factor is the threshold voltage V_T^{MW} variation of the write transistor. In fact, according to the model in (3.4), variations in this parameter have an exponential effect on the sub-V_T current through MW, which is the primary leakage mechanism affecting the RT in a 2T GC. The second most influential factor on the RT variation is t_{ox} of MW, primarily due to its effect on the sub-V_T swing (n) of MW, according to:

$$n = 1 + C_d/C_{ox} = 1 + C_d \cdot t_{ox}/\varepsilon_{ox}, \tag{3.12}$$

where C_d is the capacitance of the depletion layer; C_{ox} is the gate oxide capacitance; and ε_{ox} is equal to the vacuum permittivity multiplied by the relative permittivity of silicon dioxide. The subthreshold swing determines how well the device can be turned off, eliminating deteriorating sub-V_T leakage currents through MW. This is shown in the well-known EKV derived sub-V_T current equation (3.2).

As expected from (3.2), the dimensions of MW also significantly impact the leakage currents. As L_{MW} increases, I_{sub} decreases, lowering the rate of change of the stored data level, and thereby increasing the RT. Accordingly, we would expect to see that W_{MW} would have an equally strong impact on RT; however, Fig. 3.9 shows that the sensitivity of RT to W_{MW} is much lower. This is due to the dual effect of W_{MW} on RT. In fact, on the one hand, an increase in W_{MW} increases the sub-V_T current through MW. On the other hand, an increase in W_{MW} also increases the size of the junction capacitance contributing to C_{SN}. In addition, due to the short channel effect (SCE), variations in L_{MW} also change V_T^{MW} by as much as 20 mV within the 3σ analysis space, further contributing to the sensitivity of RT to L_{MW}.

Finally, W, L, and t_{ox} of MR affect the value of the storage capacitor, and therefore, we would expect to see a greater influence than that shown by the sensitivity analysis. Their low impact is due to the fact that the parasitic device capacitances constitute only 10% of C_{SN}, which is composed primarily of metal stack capacitance.

In conclusion, to narrow down the RT distribution and move the worst cell closer to a typical cell, it is key for process engineers to improve the threshold voltage repeatability, while variations in metal-to-metal capacitance are of a lesser concern. As a circuit designer, it is preferable to choose a CMOS technology node with low V_T variability, and/or increase both L_{MW} and W_{MW} for a higher repeatability of V_T^{MW} to achieve less RT variability.

3.6.3 Impact of Process Corner

While the previous section showed the impact of individual circuit parameters on RT, this section shortly presents the impact of the global process corner on the RT distribution. To this end, MC circuit simulations, accounting for within-die parametric variations in the TT, slow NMOS-slow PMOS (SS), and fast NMOS-fast PMOS (FF) process corners were performed. Figure 3.10 shows the simulated EDRT distributions in the TT, SS, and FF corners. Compared to a TT die, the mean EDRT decreases for a FF die and increases for a SS die. This is caused by stronger and weaker leakage current through MW in the FF and SS corners, respectively, compared to the TT corner. As expected, the EDRT variation is much lower in the FF corner than in the TT corner, and much worse in the SS corner. In fact, V_T^{MW} and, consequently, μ_{EDRT} change across process corners: in the FF corner, a smaller μ_{EDRT} leads to a smaller mean RT and less RT variance compared to the SS corner, according to (3.9).

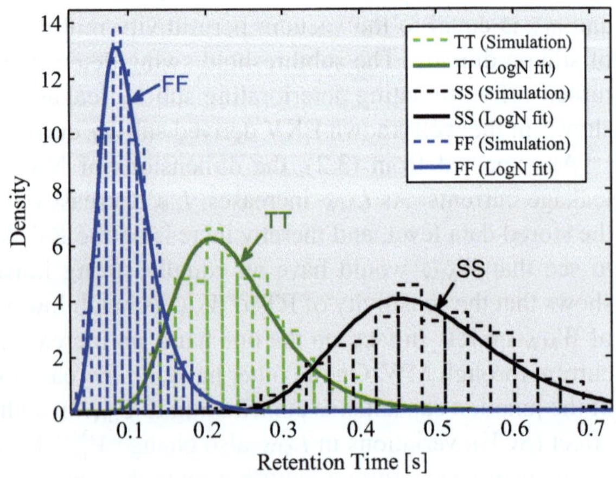

Fig. 3.10 Influence of process corners on EDRT

To conclude, the process corner has a strong impact on the RT distribution (both the mean and variance), with mean values ranging from 0.1 to 0.485 s, and RT spreads getting narrower (better) and broader (worse) in the FF and SS corners, respectively, compared to the TT corner.

3.7 Best-Practice 2T GC Design

This section describes the best-practice 2T GC design based on the insights gained from the modeling, simulations, silicon measurements, and sensitivity analysis presented before. While derived from 0.18 μm technology, these insights can also be applied to more advanced technology nodes, as long as I_{sub} is the main leakage mechanism. This is indeed often the case, especially in case of low operating voltages [2]. In cases where gate leakage is more significant, it can be added to the model. The following best-practice design guidelines maximize the RT and narrow its statistical distribution.

The mean value of the log-normal EDRT distribution is equal to $e^{\mu+\frac{\sigma^2}{2}}$. As apparent from (3.4) and Fig. 3.2, increasing the absolute value of V_T^{MW} increases the nominal EDRT value, as well as its (μ_{EDRT}) value. To this end, if several transistor families and/or V_T options are available in a given technology node, it is preferable to choose the device with the highest absolute V_T value. For example, in the herein considered 0.18 μm CMOS process, MW was implemented as an I/O device, which has a higher absolute V_T than a core device. Furthermore, as concluded in Sect. 3.6, V_T^{MW} variation is the single most influential circuit parameter responsible for the RT spread. Therefore, it is preferable to use a CMOS node and/or a device with low V_T variability, and/or increase the MW device dimensions.

Next, according to the EDRT model, μ_{EDRT} can be further increased by increasing C_{SN}, for example by taking advantage of more metal layers. Decreasing W_{MW} improves μ_{EDRT} due to lower I_{sub} through MW, but increases the standard deviation of the EDRT distribution (σ_{EDRT}), since V_T variation increases as W decreases. Increasing L_{MW} improves both μ_{EDRT} and σ_{EDRT}, due to lower sub-V_T leakage and larger channel area (lower V_T variation), respectively. Furthermore, increasing L_{MW} also increases V_T^{MW} through the short channel effect. Therefore, increasing L_{MW} has a positive impact on both μ_{EDRT} and σ_{EDRT}, and should be preferred over decreasing W_{MW} for improved RT.

The sole drawback of an increased L_{MW} is the decreased write speed, but for most embedded memory applications high write speed is less critical than high read speed. However, if high write speed is required, increasing W_{MW} by the same factor as L_{MW} is the best option to optimize the RT under a given write speed constraint.

As discussed in Sect. 3.6, the global process corner has a strong impact on both μ_{EDRT} and σ_{EDRT}. Therefore, on-chip process monitors should be used to lower the refresh rate on SS dies compared to TT and FF dies. Given the large retention time range, binning with more than two or three bins would be meaningful, as well. Chapter 5 presents a replica technique to find the optimum refresh rate for each die in order to reduce refresh power and enable low-power GC-eDRAM designs.

Finally, since EDRT depends on the read frequency, it does not only depend on the GC parameters, but also on the readout circuits. In general, a circuit resolving a smaller voltage difference in a shorter time leads to longer EDRT.

3.8 Conclusions

This chapter introduced an analytical model for the statistical per-cell retention time distribution of GC-eDRAM. While all circuit parameters follow a Gaussian distribution, it was shown that the per-cell retention time follows a log-normal distribution. The simple analytical retention time distribution model is highly precise when compared to Monte Carlo circuit simulation and silicon measurement results. In fact, modeling errors are less than 6%.

A PB-DOE sensitivity analysis unveiled that the threshold voltage of the write transistor is the sole most dominant parameter whose uncertainty leads to large per-cell retention time spread. Variability in all other circuit parameters, including the storage node capacitor (built from the metal stack available in digital CMOS technologies), is responsible for only a small percentage of the total retention time variability.

While the coefficients of the analytical models for the nominal value of the effective retention time as well as for its statistical distribution can be derived from primary GC circuit parameters, it is also possible to extract the coefficient from curve fitting based on a small number of circuit simulations or silicon measurements. The coefficients extracted through curve fitting were shown to precisely match the original coefficients of the analytical model. In conclusion, fitting a log-normal distribution to a small set of simulated or measured retention time data is a convenient shortcut to obtain a precise analytical model of the retention time distribution, closely tracking a more fundamental analytical model derived from primary circuit parameters.

The presented analytical model is useful to estimate the worst-case retention time of large memory arrays in a short time, whereas the collection of a sufficient number of MC samples would be time-consuming.

The model was proven through silicon measurements in a mature 0.18 μm CMOS process node. In addition, Chap. 6 shows that the model continues to hold up in advanced process nodes, such as 28 nm CMOS.

References

1. Burmaster DE, Hull DA (1997) Using lognormal distributions and lognormal probability plots in probabilistic risk assessments. Hum Ecol Risk Assess 3(2):235–255
2. Butzen PF, Ribas RP (2006) Leakage current in sub-micrometer CMOS gates. Universidade Federal do Rio Grande do Sul, pp 1–28

3. Chang MT, Rosenfeld P, Lu SL, Jacob B (2013) Technology comparison for large last-level caches (L^3Cs): low-leakage SRAM, low write-energy STT-RAM, and refresh-optimized eDRAM. In: IEEE International symposium on high performance computer architecture (HPCA), pp 143–154
4. Cheng Y, Chan M, Hui K, Jeng Mc, Liu Z, Huang J, Chen K, Chen J, Tu R, Ko PK, et al (1996) BSIM3v3 manual. University of California, Berkeley
5. Chun KC, Jain P, Lee JH, Kim C (2011) A 3T gain cell embedded DRAM utilizing preferential boosting for high density and low power on-die caches. IEEE J Solid State Circuits 46(6):1495–1505
6. Chun KC, Jain P, Kim TH, Kim C (2012) A 667 MHz logic-compatible embedded DRAM featuring an asymmetric 2T gain cell for high speed on-die caches. IEEE J Solid State Circuits 47(2):547–559. doi:10.1109/JSSC.2011.2168729
7. Chun KC, Zhang W, Jain P, Kim C (2012) A 2T1C embedded DRAM macro with no boosted supplies featuring a 7T SRAM based repair and a cell storage monitor. IEEE J Solid State Circuits 47(10):2517–2526
8. Do AT, Yi H, Yeo KS, Kim T (2012) Retention time characterization and optimization of logic-compatible embedded DRAM cells. In: Proceedings of the IEEE Asia symposium on quality electronic design (ASQED), pp 29–34
9. Ghosh S (2014) Modeling of retention time for high-speed embedded dynamic random access memories. IEEE Trans Circuits Syst I 61(9):2596–2604
10. Giterman R, Teman A, Meinerzhagen P, Atias L, Burg A, Fish A (2016) Single-supply 3T gain-cell for low-voltage low-power applications. IEEE Trans Very Large Scale Integr (VLSI) Syst 24(1):358–362
11. Karakonstantis G, Roth C, Benkeser C, Burg A (2012) On the exploitation of the inherent error resilience of wireless systems under unreliable silicon. In: Proceedings of the ACM/EDAC/IEEE design automation conference (DAC), pp 510–515
12. Lin DK, Draper NR (1993) Generating alias relationships for two-level Plackett and Burman designs. Comput Stat Data Anal 15(2):147–157
13. Mande S, Chandorkar AN, Hsaio C, Huang K, Sheu YM, Liu S (2009) A novel approach to link process parameters to BSIM model parameters. IEEE Trans Semicond Manuf 22(4):544–551
14. Mande SS, Chandorkar AN, Iwai H (2013) Computationally efficient methodology for statistical characterization and yield estimation due to inter-and intra-die process variations. In: Proceedings of the IEEE Asia symposium on quality electronic design (ASQED), pp 287–294
15. Meinerzhagen P, Teman A, Mordakhay A, Burg A, Fish A (2012) A sub-VT 2T gain-cell memory for biomedical applications. In: Proceedings of the IEEE subthreshold microelectronics conference (SubVT), pp 1–3. doi:10.1109/SubVT.2012.6404318
16. Meinerzhagen P, Teman A, Burg A, Fish A (2013) Impact of body biasing on the retention time of gain-cell memories. IET J Eng 1:1–4
17. Meinerzhagen P, Teman A, Giterman R, Burg A, Fish A (2013) Exploration of sub-VT and near-VT 2T gain-cell memories for ultra-low power applications under technology scaling. J Low Power Electron Appl 3(2):54–72. doi:10.3390/jlpea3020054
18. Mokhov A, Rykunov M, Sokolov D, Yakovlev A (2014) Design of processors with reconfigurable microarchitecture. J Low Power Electron Appl 4(1):26–43
19. Park YS, Blaauw D, Sylvester D, Zhang Z (2012) A 1.6-mm^2 38-mW 1.5-Gb/s LDPC decoder enabled by refresh-free embedded DRAM. In: Proceedings of the IEEE symposium on VLSI circuits (VLSIC), pp 114–115
20. Shauly EN (2012) CMOS leakage and power reduction in transistors and circuits: process and layout considerations. J Low Power Electron Appl 2(1):1–29
21. Somasekhar D, Ye Y, Aseron P, Lu SL, Khellah M, Howard J, Ruhl G, Karnik T, Borkar S, De V, Keshavarzi A (2009) 2 GHz 2 Mb 2T gain cell memory macro with 128 GBytes/sec bandwidth in a 65 nm logic process technology. IEEE J Solid State Circuits 44(1):174–185
22. Teman A, Meinerzhagen P, Burg A, Fish A (2012) Review and classification of gain cell eDRAM implementations. In: Proceedings of the IEEE convention of electrical and electronics engineers in Israel (IEEEI), pp 1–5

23. Teman A, Meinerzhagen P, Giterman R, Fish A, Burg A (2014) Replica technique for adaptive refresh timing of gain-cell embedded DRAM. IEEE Trans Circuits Syst II 61(4):259–263
24. Teman A, Karakonstantis G, Giterman R, Meinerzhagen P, Burg A (2015) Energy versus data integrity trade-offs in embedded high-density logic compatible dynamic memories. In: Proceedings of the ACM design, automation & test in Europe conference & exhibition (DATE), DATE '15, pp 489–494
25. Wilkerson C, Alameldeen AR, Chishti Z, Wu W, Somasekhar D, Lu Sl (2010) Reducing cache power with low-cost, multi-bit error-correcting codes. ACM SIGARCH Comput Architect News 38(3):83–93
26. Xie YF, Cheng K, Lin YY (2012) A logic 2T gain cell eDRAM with enhanced retention and fast write scheme. In: Proceedings of the IEEE international conference on solid-state and integrated circuit technology (ICSICT), pp 1–3
27. Zhang W, Chun KC, Kim CH (2010) Variation aware performance analysis of gain cell embedded DRAMs. In: Proceedings of the ACM/IEEE international symposium on low power electronics and design (ISLPED), pp 19–24
28. Zhang X, Leomant S, Lau KL, Bermak A (2011) A compact digital pixel sensor (DPS) using 2T-DRAM. J Low Power Electron Appl 1(1):77–96

Chapter 4
Conventional GC-eDRAMs Scaled to Near-Threshold Voltage (NTV)

With contributions by Rashid Iqbal

4.1 Introduction

As seen in Chap. 2, almost all previous works on GC-eDRAM considered operation at nominal supply voltage for high speed performance and high memory bandwidth. As opposed to this, the present chapter investigates the impact of voltage scaling on the retention time and power consumption of a 2-transistor (2T)-bitcell GC-eDRAM. Targeting near-threshold computing (NTC) [6] systems (see middle column of Table 1.1 in Chap. 1) which are characterized by low power consumption at still relatively high speed performance, this chapter investigates the limit of voltage scaling for GC-eDRAM such that all operations still rely on on-currents of the inherent transistors. Relying on subthreshold conduction for active operations will be addressed later on in Chaps. 6 and 7). The voltage limit for the main supply which still ensures fast circuit operation thanks to on-currents is derived for both the case of using an underdrive voltage for the write word-line (WWL) and for the case of using a single main supply voltage for the entire GC-eDRAM macrocell. Interestingly, the retention time can be increased when scaling down the supply voltage for given memory access statistics and a given write bit-line (WBL) control scheme. Moreover, for a given supply voltage, the retention time can be further increased by controlling the WBL to a voltage level between the two supply rails (V_{DD} and V_{SS}) during idle and read states (which, however, entails an overhead for voltage generation). These two concepts are proved by means of Spectre simulation of a GC-eDRAM macrocell implemented in 180 nm CMOS technology and operated at only 40% of the nominal supply voltage. In order to maintain high memory bandwidth even for reduced operating frequencies at scaled voltages, it is shown that a 2T-bitcell GC-eDRAM macrocell can easily be implemented as a two-port memory at a negligible area overhead compared to a single-port memory implementation.

P. Meinerzhagen et al., *Gain-Cell Embedded DRAMs for Low-Power VLSI Systems-on-Chip*, DOI 10.1007/978-3-319-60402-2_4

The various embedded memory technologies introduced in Sect. 1.3, namely static random-access memory (SRAM), embedded DRAM (eDRAM), and latch/flip-flop arrays, are quickly reviewed in the light of supply voltage scaling in the following. Specially designed SRAM macrocells can operate reliably at scaled supply voltages at the price of relatively large 8-transistor (8T) [2], 10T [1], or even 14T [7] bitcells, and/or dedicated low-voltage read and write assist circuits [11, 14, 17]. Alternatively, synthesized latch arrays or flip-flop arrays are a more straightforward approach to reliable low-voltage storage arrays than SRAMs but have an even larger area cost for storage capacities higher than a few kb [13]. In conventional 1-transistor-1-capacitor (1T-1C) embedded DRAM (eDRAM), the offset voltage of the sense amplifier limits voltage downscaling, unless dedicated offset cancellation techniques are used [8]. Another major obstacle in low-voltage 1T-1C eDRAM is the degradation of the data retention time, which requires power-consuming refresh operations more frequently [8]. Moreover, as expatiated on in Chap. 2, conventional 1T-1C eDRAMs require special process options to build high-density 3D capacitors (stacked or trench capacitors), which adds cost to standard digital CMOS technologies. As a further attractive option for building embedded storage arrays operated at scaled voltages, gain-cells (GCs) are smaller than any SRAM bitcell, latch, and flip-flop, while they are fully compatible with standard digital CMOS technologies. While most previous works available in literature promote GC-eDRAM as denser successor of SRAM for on-die caches in high-end processors [5, 16] (see Chap. 2 for details), only a small number of works investigate GC-eDRAM operation at scaled voltages: (1) a dual threshold-voltage (dual-V_T) GC array [12] is operated at a fraction of the nominal supply voltage; the circuit increases the retention time by using a high threshold voltage (high-V_T) write transistor (MW); and (2) another storage macro based on a boosted 3-transistor (3T) GC [4] is operable in a supply voltage range from 1.2 V down to 0.7 V and uses preferential storage node (SN) boosting while reading to increase the retention time and the read speed.

Previously reported GC-eDRAM macrocells are not clearly classified as either single-port or two-port implementations. Furthermore, while previous work on GC-eDRAMs targets a given supply voltage (or supply voltage range) and presents dedicated techniques to increase the retention time, the impact of supply voltage scaling on the retention time has not been systematically investigated yet. Moreover, previous publications do not clearly state the assumed write access statistics for the measurement of the retention time, while frequent write accesses may in fact significantly degrade the retention time.

Therefore, the remainder of this chapter reviews in detail why GCs are inherently suitable for two-port memory implementations with a negligible area-overhead compared to single-port implementations. Next, the limit to supply voltage scaling in conventional 2T-bitcell GC-eDRAM in the occurrence of process parameter variation is discussed, avoiding relying on subthreshold conduction to achieve the medium speed performance required by NTC systems. The impact of supply voltage downscaling on the retention time under well-defined memory access statistics is then investigated, allowing for finding the optimum supply voltage for highest retention time and lowest power consumption. Finally, a simple technique to further improve the retention time at any given supply voltage is presented.

4.2 2T GC, Array, and Macrocell Optimized for NTV Operation

4.2.1 2T Two-Port GC and Array Architecture

Concurrent read and write access is an effective method for achieving high memory bandwidth [10]. Two-port memories have a separate read and write port to enable such access. In conventional 1T-1C DRAM and conventional SRAM, the same word-lines (WLs) and bit-lines (BLs) are used for both the read and the write operation; enabling two-port operation is nontrivial and requires additional hardware in each bitcell. As opposed to this, GCs are inherently well suited for two-port operation, as they already have a separate read port consisting of the read word-line (RWL) terminal and the read bit-line (RBL) terminal, as well as a separate write port consisting of the write word-line (WWL) terminal and the write bit-line (WBL) terminal, as shown in Fig. 4.1. It is therefore straightforward to enable two-port operation in GC-eDRAMs and benefit from the resulting high memory bandwidth, even for operation in the near-V_T domain.

In the two-port memory architecture adopted in this chapter, there are two address decoders: one for the write address, and another one for the read address. A single-port implementation would save one address decoder, but it would require additional logic circuits—comparable in size to a single decoder—to distribute the decoded address to either the write port or the read port, while silencing the other port.

Other than the explicit two-port configuration, the memory architecture serving as a basis for the herein presented voltage scaling analysis is mostly adopted from [12]. As shown in Fig. 4.2, the storage array consists of 32 rows and 64 columns. Moreover, the conventional sense amplifiers are replaced with simple sense inverters to improve the area-efficiency [12]. To allow for conclusions as

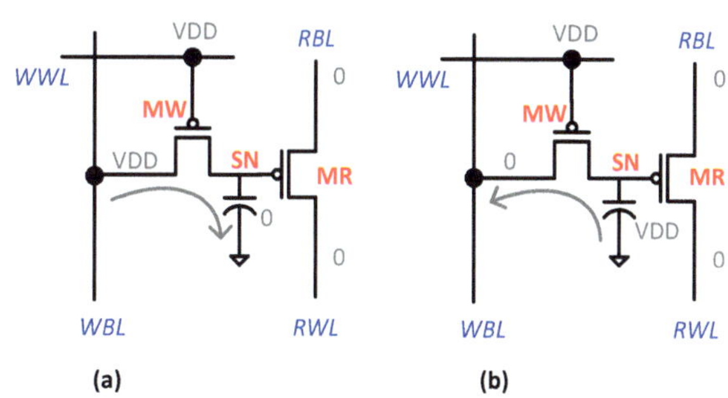

Fig. 4.1 2-PMOS gain-cell; worst write bit-line (WBL) state for retention of (**a**) logic "0" and (**b**) logic "1"

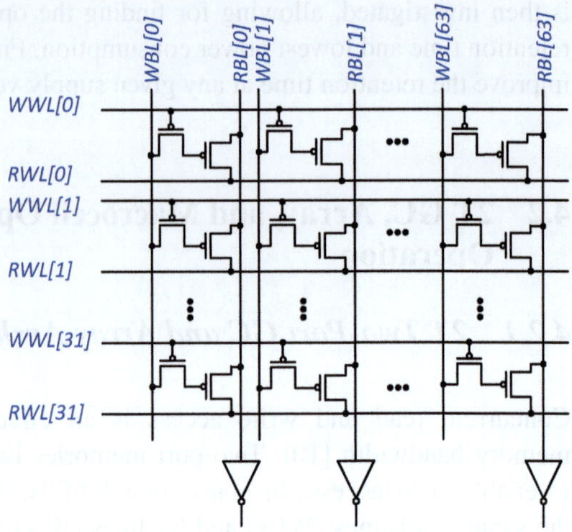

Fig. 4.2 2T-bitcell GC-eDRAM storage array with area-efficient sense inverters

general as possible, the basic 2-PMOS GC with regular threshold voltage (regular-V_T) transistors from [16] is adopted in this work, as the high-V_T transistors used in [12] might not be available in all CMOS technology nodes. Notice, however, that high-V_T transistors can reduce subthreshold conduction by more than two orders of magnitude compared to regular-V_T transistors [12], and therefore allow for considerably longer retention times, as will be discussed in detail in Chaps. 5 and 6.

4.2.2 Operation Principle

4.2.2.1 Hold, Write, and Read Operations

In each 2T GC, data is stored in the form of charge on the SN capacitor, which is formed by the gate capacitance of the storage/read transistor (MR) as well as the junction capacitance of MW and wire (metal and via) parasitic capacitances. The parasitic SN capacitor is shown explicitly in Fig. 4.1, even though it is not an explicit circuit component (unless dedicated metal finger capacitors are added to the bitcell). During a write operation, MW of the selected GC is turned on to transfer the new data level from the WBL to the SN. To allow the transfer of a clean logic "0," an underdrive voltage of −500 mV is applied to the selected WWL. At the beginning of a read operation, all RBLs are discharged to ground (V_{SS}). Next, the selected RWL is pulled high to V_{DD}. If a GC stores a logic "1," its MR remains off and the connected RBL ideally remains at V_{SS} (leakage current may slowly charge RBL). However, if the GC stores a logic "0," the RBL starts to charge quickly through MR. The sense inverter must switch before RBL is charged to the threshold voltage of MR (V_T^{MR}), as at this time read transistors MR in unselected cells storing logic "0" turn on, which provides a current path to V_{SS} and prevents a further voltage rise on the RBL.

4.2.2.2 Limit to Supply Voltage Scaling for Fast Access

The minimum supply voltage for reasonably fast memory access is determined by the ability of writing, holding, and reading two distinct data levels while not relying on subthreshold conduction for active circuit operation. Considering the 2-PMOS GC and avoiding any underdrive voltage, MW can easily transfer a high voltage level equal to V_{DD} to the SN. However, the lowest data level which can be transferred in a reasonable time, i.e., not relying on subthreshold conduction, is equal to the threshold-voltage of MW ($V_{\mathrm{T}}^{\mathrm{MW}}$). When turning off MW, charge injection and clock feedthrough rise the voltage on the SN (V_{SN}) by ΔV_{SN}; the amount of voltage rise depends on the SN capacitance, the voltage level being transferred, and several other factors. After writing a logic "0" level, $V_{\mathrm{SN}} = V_{\mathrm{T}}^{\mathrm{MW}} + \Delta V_{\mathrm{SN}}$. Holding a data level on the SN during a small amount of time is possible regardless of V_{DD}. To tell a logic "0" from a logic "1" at the time of reading, V_{SN} must be smaller than $V_{\mathrm{DD}} - V_{\mathrm{T}}^{\mathrm{MR}}$ in order to still be able to turn on the RT:

$$V_{\mathrm{T}}^{\mathrm{MW}} + \Delta V_{\mathrm{SN}} < V_{\mathrm{DD}} - V_{\mathrm{T}}^{\mathrm{MR}} \tag{4.1}$$

Equation (4.1) is rearranged to show the lower limit for V_{DD}:

$$V_{\mathrm{T}}^{\mathrm{MW}} + V_{\mathrm{T}}^{\mathrm{MR}} + \Delta V_{\mathrm{SN}} < V_{\mathrm{DD}} \tag{4.2}$$

To account for process parameter variations (both die-to-die and within-die variations), Eq. (4.2) is rewritten as follows, where $\mu(X)$ and $\sigma(X)$ denote the mean and the standard deviation of the random variable X.

$$\left(\mu(V_{\mathrm{T}}^{\mathrm{MW}}) + N\sigma(V_{\mathrm{T}}^{\mathrm{MW}})\right) + \left(\mu(V_{\mathrm{T}}^{\mathrm{MR}}) + N\sigma(V_{\mathrm{T}}^{\mathrm{MR}})\right) + \Delta V_{\mathrm{SN}} < V_{\mathrm{DD}} \tag{4.3}$$

The parameter N is chosen depending on the desired yield. For small GC-eDRAM arrays of several kb, $N = 3$ is reasonable.

Assuming a WWL underdrive, a clean V_{SS} level can be transferred to the SN, and V_{DD} can be further reduced, with its lower limit now given by:

$$\left(\mu(V_{\mathrm{T}}^{\mathrm{MR}}) + N\sigma(V_{\mathrm{T}}^{\mathrm{MR}})\right) + \Delta V_{\mathrm{SN}} < V_{\mathrm{DD}} \tag{4.4}$$

It is usually beneficial in terms of energy efficiency to have a WWL underdrive, as most parts of the circuit can be operated from a lower V_{DD}, while the underdrive voltage is only applied to the write address decoder and the WWL drivers.

In the current case, using an underdrive voltage of $-500\,\mathrm{mV}$, and with $\mu(V_{\mathrm{T}}^{\mathrm{MR}}) = 500\,\mathrm{mV}$, $\sigma(V_{\mathrm{T}}^{\mathrm{MR}}) = 25\,\mathrm{mV}$, $N = 3$, $\Delta V_{\mathrm{SN}} \approx 100\,\mathrm{mV}$ (extracted from circuit simulations in 180 nm CMOS technology), and a small margin for uncertainty in ΔV_{SN}, the lowest V_{DD} for reliable operation and reasonable yield is 700 mV, which is only 40% of the nominal V_{DD} (1.8 V).

4.3 Impact of Voltage Scaling on GC-eDRAM Retention Time

Low-voltage low-to-medium speed VLSI systems-on-chip (such as complex sensor nodes) are best implemented in older, low-leakage CMOS technology nodes (such as 180 nm) to minimize energy dissipation, especially if leakage-reduction techniques such as power gating are applied [15]. The considered GC-eDRAM array is therefore implemented in a commercial 180 nm CMOS technology. Among many leakage mechanisms, the subthreshold conduction of MW is clearly the dominant leakage mechanism corrupting the stored data. The subthreshold conduction and consequently the data retention time strongly depend on the voltage level encountered on the WBL, denoted by V_{WBL}.

Assuming that a GC has just been written and is now holding its data, there are two possible scenarios:

1. Further write operations are performed to GCs on the same WBL, which means that V_{WBL} is data-dependent and cannot be controlled.
2. The memory macrocell remains in idle state (no data accesses) or only read accesses are performed. During idle and read states, V_{WBL} can be controlled to any desired voltage level to minimize subthreshold conduction through MW.

Figure 4.1 shows the *worst-case access* scenario in terms of retention time where the opposite data level is permanently written to GCs on the same WBL after writing a given data level to the first GC. The *retention mode* scenario presumes an application where a relatively small storage array (with only few GCs per WBL) is fully written in a negligibly short time, whereafter the memory macrocell is kept in idle or read states and the WBL can be controlled to either V_{DD} or V_{SS}. Under the same retention mode scenario, the potential of controlling the WBL to a voltage level between the supply rails (V_{DD} and V_{SS}) will be evaluated, as well.

4.3.1 *Worst-Case Access*

Assuming the worst-case access scenario where V_{WBL} is permanently opposite to the stored data level, this chapter defines the retention time for a logic “0” (“1”), denoted by t_{ret0} (t_{ret1}), as the time it takes for V_{SN} to rise (fall) to $V_{DD} - V_T^{MR}$ (see Chap. 3 for other definitions of retention time). At nominal V_{DD}, t_{ret1} is longer than t_{ret0}: the more the logic “1” voltage level decays, the more positive the gate-to-source voltage V_{GS} and the higher the reverse body biasing (RBB) of MW, both suppressing the subthreshold conduction harder [16].

As illustrated in Fig. 4.3, when V_{DD} is gradually scaled down, the storage range for a logic “0,” given by $V_{DD} - V_T^{MR}$ (if neglecting charge sharing and clock feedthrough for simplicity), becomes smaller, while the storage range for a logic “1,”

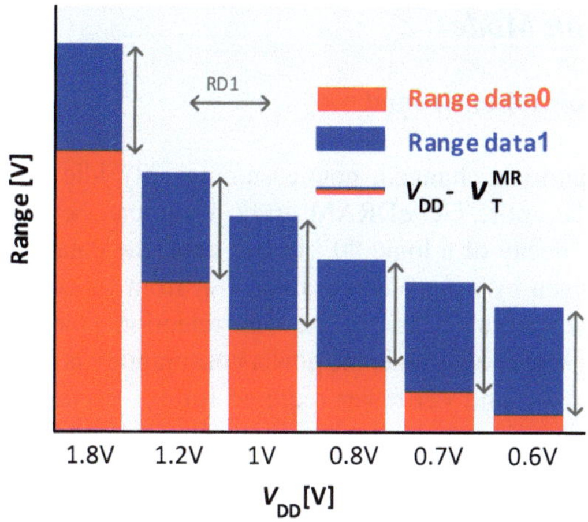

Fig. 4.3 Storage ranges (voltage ranges) for data "0" and "1" versus main supply voltage V_{DD}

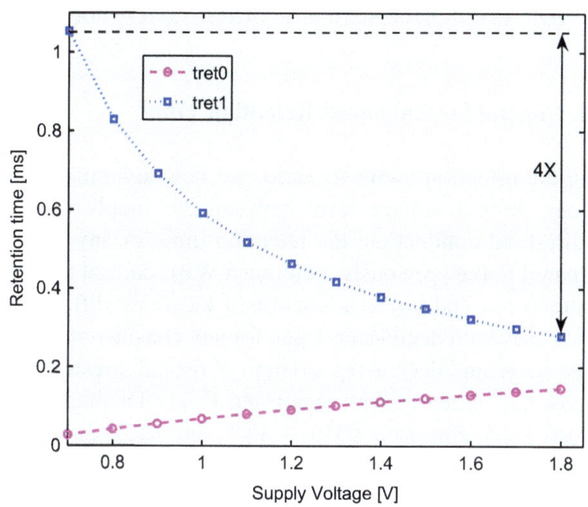

Fig. 4.4 Retention time vs. V_{DD} for worst-case WBL state (always opposite to stored data)

given by V_T^{MR}, remains unchanged. At the same time, when V_{DD} is scaled down, the subthreshold conduction of MW becomes weaker due to its exponential dependence on V_{GS} and the drain-to-source voltage V_{DS}.

Consequently, the nominal value of t_{ret1} increases with decreasing V_{DD}, as shown by the Spectre circuit simulation results in Fig. 4.4. However, Fig. 4.4 also shows that t_{ret0} decreases with decreasing V_{DD}, as the always smaller storage range has the higher impact than the decreasing strength of the subthreshold conduction.

4.3.2 *Retention Mode*

4.3.2.1 WBL Control to Ground

If the access scenario is changed, now assuming only idle and read states after initially writing the entire GC-eDRAM array, V_{WBL} can be controlled to V_{SS}, in order to avoid the decay of a logic "0." In this case, the data retention time of the storage array is given by t_{ret1}. When scaling V_{DD} from its nominal value of 1.8 V down to 0.7 V, the nominal value of the data retention time increases by 4×, as shown in Fig. 4.4. At the same time, the power consumption is considerably reduced, due to (1) lower V_{DD}, and (2) fewer required refresh cycles. These results were extracted from basic circuit simulations in a mature 180 nm CMOS technology, characterized by little parametric variations. For more advanced CMOS nodes, discussed in Chaps. 6 and 7, it is key to account for global and local parametric variations when estimating the retention time by means of circuit simulation. In summary, if the GC-eDRAM is kept in idle/hold or read states after an initial write access, supply voltage scaling improves both retention time and energy efficiency, at least for GC-eDRAM implementations in mature CMOS nodes.

4.3.2.2 WBL Control for Enhanced Retention Time

Still presuming the retention mode scenario, but now assuming that V_{WBL} can be controlled to any desired voltage level between the supply rails V_{DD} and V_{SS}[1] to reduce subthreshold conduction, the retention time for any V_{DD} can be further increased compared to the previously mentioned WBL control to V_{SS}.

Figure 4.5 shows t_{ret1} and t_{ret0} as a function of V_{WBL}, for different values of V_{DD}. Clearly, t_{ret0} increases with decreasing V_{WBL} for any considered V_{DD} value, due to a constant storage range and decreasing strength of the subthreshold conduction. For the same reasons, t_{ret1} increases with increasing V_{WBL}. The highest retention times are reached when V_{WBL} approaches $V_{DD} - V_T^{MR}$, and t_{ret1} (t_{ret0}) becomes infinitely long for V_{WBL} higher (lower) than $V_{DD} - V_T^{MR}$. However, the slopes in this region are very steep, so that any noise on V_{WBL} can considerably degrade the retention time. At $V_{DD} = 700\,\text{mV}$, choosing $V_{WBL} = 200\,\text{mV}$, a retention time of 3.3 ms is achieved, corresponding to a 3.3× improvement compared the case where V_{WBL} is controlled to V_{SS}.

---

[1]Of course, controlling V_{WBL} to a voltage level between the main supply rails requires additional circuits (DC–DC voltage converters) whose use can only be justified for large GC-eDRAM arrays.

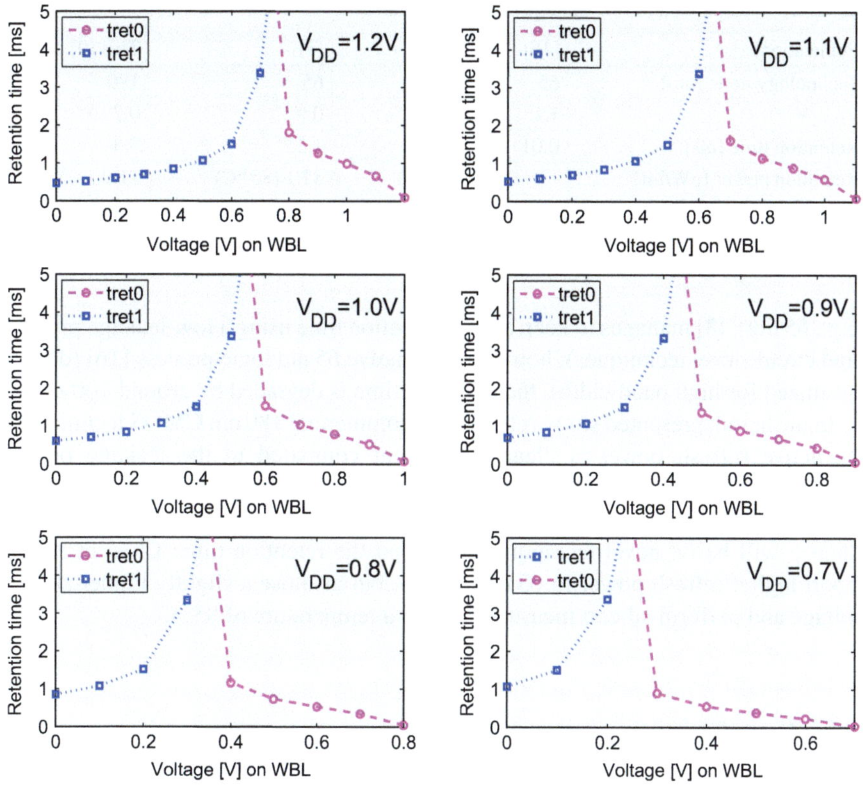

Fig. 4.5 WBL control to intermediate voltage levels for enhanced retention time

4.4 Macrocell Implementation Results

This section presents simulation results of the 32 × 64 2T GC-eDRAM macrocell with WBL control implemented in 180 nm CMOS technology, and compares it with the state-of-the-art GC-eDRAM macrocells in literature. In the retention mode, an overall improvement of 13.2× in retention time and a considerable reduction in power consumption are obtained by supply voltage scaling and the above-mentioned WBL control technique. The active refresh power of the presented 2 kb GC-eDRAM macrocell is 10.8 pW/bit, while the leakage power is 1.1 pW/bit, amounting to a total data retention power of 11.9 pW/bit.

Table 4.1 compares this work to a selection of GC-eDRAMs available in literature [3, 12, 16]. All retention time and retention power values are given for a temperature of 25 °C, unless stated otherwise.

For the same technology node (i.e., 180 nm), Table 4.1 shows the effectiveness of a high-V_T write transistor (MW) [12] (if available and financially interesting) to improve the retention time by around 100×. For smaller technology nodes

Table 4.1 Comparison of low-voltage GC-eDRAM arrays

Publication	[16]	[12]	[3]	This [9]
Technology node [nm]	65	180	65	180
V_{DD} [V]	1.1	0.75	0.9	0.7
Retention time [ms]	0.01	306[a]	1.25[b]	3.3
Retention power [pW/bit]	-	0.662	87.1 (85 °C)	11.9

[a]High-V_T transistor reduces leakage by more than two orders of magnitude [12]
[b]Low-leakage CMOS technology

(e.g., 65 nm), [3] manages to keep a good retention time using a low-leakage process (and circuit-level techniques); however, in a native 65 nm logic process [16] (design optimized for high bandwidth), the retention time is degraded by around 100×.

In the herein presented study relying on a commercial 180 nm CMOS technology, the active refresh power is clearly dominant compared to the leakage power, meaning that any effort to increase the retention time also significantly reduces the total data retention power (see Table 4.1). Therefore, the focus of the following Chap. 5 will be on novel techniques to extend the retention time. Chun et al. [3] report higher refresh power in 65 nm CMOS, but also use a slightly higher supply voltage and perform silicon measurements at a temperature of 85 °C.

4.5 Conclusions

GC-eDRAMs are an interesting alternative to SRAM macrocells in low-power/low-voltage (near-V_T) VLSI SoCs and microprocessors. GCs are inherently suitable for the construction of two-port memories, as opposed to SRAM and conventional eDRAM, and can therefore achieve high memory bandwidth even at near-V_T voltages. A conventional 2T (2-PMOS)-bitcell GC-eDRAM can be reliably operated at low supply voltages close to the transistor's threshold voltage if a few critical circuit nodes, namely the WWLs, receive an underdrive voltage.

For a conventional 2T GC implemented in 180 nm CMOS technology, the data retention time improves by 4× when scaling down the supply voltage from 1.8 to 0.7 V, provided that write access is unfrequent and short. In addition to this, another 3.3× improvement in retention time is achieved by controlling the voltage on the WBL to a value between the supply rails (V_{DD} and V_{SS}) during idle and read states. This overall 13.2× improvement in retention time combined with operation at less than 40% of the technology's nominal V_{DD} leads to a data retention power of 11.9 pW/bit. The data retention power was found to be dominated by active refresh power, while leakage power plays only a minor role. Therefore, the next chapter (Chap. 5) presents several novel techniques to further enhance the retention time of near-V_T GC-eDRAMs for reduced data retention power.

References

1. Calhoun B, Chandrakasan A (2007) A 256-kb 65-nm sub-threshold SRAM design for ultra-low-voltage operation. IEEE J Solid State Circuits 42(3):680–688
2. Chiu YW, Lin JY, Tu MH, Jou SJ, Chuang CT (2011) 8T single-ended sub-threshold SRAM with cross-point data-aware write operation. In: Proceedings of the IEEE/ACM international symposium on low-power electronics and design (ISLPED), pp 169–174
3. Chun KC, Jain P, Lee JH, Kim C (2009) A sub-0.9V logic-compatible embedded DRAM with boosted 3T gain cell, regulated bit-line write scheme and PVT-tracking read reference bias. In: Proceedings of the IEEE symposium on VLSI circuits (VLSIC), pp 134–135
4. Chun KC, Jain P, Kim C (2010) Logic-compatible embedded DRAM design for memory intensive low power systems. In: Proceedings of the IEEE international symposium on circuits and systems (ISCAS), pp 277–280
5. Chun KC, Jain P, Lee JH, Kim C (2011) A 3T gain cell embedded DRAM utilizing preferential boosting for high density and low power on-die caches. IEEE J Solid State Circuits 46(6):1495–1505
6. Dreslinski R, Wieckowski M, Blaauw D, Sylvester D, Mudge T (2010) Near-threshold computing: Reclaiming moore's law through energy efficient integrated circuits. Proc IEEE 98(2):253–266
7. Hanson S, Seok M, Lin YS, Foo Z, Kim D, Lee Y, Liu N, Sylvester D, Blaauw D (2009) A low-voltage processor for sensing applications with picowatt standby mode. IEEE J Solid State Circuits 44(4):1145–1155
8. Hong S, Kim S, Wee JK, Lee S (2002) Low-votage DRAM sensing scheme with offset-cancellation sense amplifier. IEEE J Solid State Circuits 37(10):1356–1360
9. Iqbal R, Meinerzhagen P, Burg A (2012) Two-port low-power gain-cell storage array: voltage scaling and retention time. In: Proceedings of the IEEE international symposium on circuits and systems (ISCAS), pp 2469–2472
10. Kaku M, Iwai H, Nagai T, Wada M, Suzuki A, Takai T, Itoga N, Miyazaki T, Iwai T, Takenaka H, Hojo T, Miyano S, Otsuka N (2008) An 833MHz pseudo-two-port embedded DRAM for graphics applications. In: Proceedings of the IEEE international solid-state circuits conference (ISSCC), pp 276–613. doi:10.1109/ISSCC.2008.4523164
11. Karl E, Wang Y, Ng YG, Guo Z, Hamzaoglu F, Bhattacharya U, Zhang K, Mistry K, Bohr M (2012) A 4.6 GHz 162 Mb SRAM design in 22nm tri-gate CMOS technology with integrated active VMIN-enhancing assist circuitry. In: Proceedings of the IEEE international solid-state circuits conference (ISSCC), pp 230–232
12. Lee Y, Chen MT, Park J, Sylvester D, Blaauw D (2010) A 5.42nW/kB retention power logic-compatible embedded DRAM with 2T dual-VT gain cell for low power sensing applications. In: Proceedings of the IEEE Asian solid state circuits conference (A-SSCC), pp 1–4
13. Meinerzhagen P, Sherazi S, Burg A, Rodrigues J (2011) Benchmarking of standard-cell based memories in the sub-VT domain in 65-nm CMOS technology. IEEE J Emerging Sel Top Circuits Syst 1(2):173–182
14. Raychowdhury A, Geuskens B, Kulkarni J, Tschanz J, Bowman K, Karnik T, Lu SL, De V, Khellah M (2010) PVT-and-aging adaptive wordline boosting for 8T SRAM power reduction. In: Proceedings of the IEEE international solid-state circuits conference (ISSCC), pp 352–353
15. Seok M, Sylvester D, Blaauw D (2008) Optimal technology selection for minimizing energy and variability in low voltage applications. In: Proceedings of the ACM/IEEE international symposium on low power electronics and design (ISLPED), pp 9–14
16. Somasekhar D, Ye Y, Aseron P, Lu SL, Khellah M, Howard J, Ruhl G, Karnik T, Borkar S, De V, Keshavarzi A (2008) 2 GHz 2 Mb 2T gain-cell memory macro with 128 GB/s bandwidth in a 65 nm logic process. In: Proceedings of the IEEE international solid-state circuits conference (ISSCC), pp 274–613
17. Zhang K, Bhattacharya U, Chen Z, Hamzaoglu F, Murray D, Vallepalli N, Wang Y, Zheng B, Bohr M (2006) A 3-GHz 70-Mb SRAM in 65-nm CMOS technology with integrated column-based dynamic power supply. IEEE J Solid State Circuits 41(1):146–151

Chapter 5
Novel Bitcells and Assist Techniques for NTV GC-eDRAMs

5.1 Introduction

Chapter 4 showed that supply voltage scaling to the near-threshold (near-V_T) domain is beneficial to improve the nominal retention time of 2T-bitcell GC-eDRAMs, provided that write access occurs only seldom and that the write bit-lines (WBLs) can therefore be driven to a beneficial voltage level during the majority of the time. In this chapter, three novel bitcell circuit and assist techniques to further enhance the retention time of near-V_T GC-eDRAMs are presented.

The first technique, presented in Sect. 5.2, introduces a novel 3T GC design, characterized by a full NMOS plus PMOS transmission gate as write port. This topology overcomes the drawback of traditional GC-eDRAM implementations that require boosted control signals in order to write full voltage levels to the SN in order to reduce the refresh rate and shorten access times. These boosted voltage levels require either an extra power supply or on-chip charge pumps, as well as nontrivial level shifting and tolerance of high voltage swings.

This chapter presents a novel 3T GC-eDRAM bitcell that operates with a single supply voltage and provides superior write capability compared to conventional GC topologies. This is achieved by using a full NMOS plus PMOS transmission gate in the write port and a single NMOS as read transistor. A 2 kb memory macro containing the proposed 3T GC was designed and fabricated in a mature 0.18 µm CMOS process. The test array is powered with a single supply of 900 mV, demonstrating a 0.8 ms worst-case retention time, a 1.3 ns write-access time, and 2.4 pW/bit of retention power. The proposed topology provides a bitcell area reduction of 43% compared to a redrawn 6T SRAM and an overall macro area reduction of 67%.

The second technique, expatiated on in Sect. 5.3, is reverse body biasing (RBB) in order to suppress the subthreshold conduction of the write transistor (MW), thereby improving the GC-eDRAM retention time. RBB has previously been applied to conventional 1T-1C DRAM, and, in this context, is also referred to as

P. Meinerzhagen et al., *Gain-Cell Embedded DRAMs for Low-Power VLSI Systems-on-Chip*, DOI 10.1007/978-3-319-60402-2_5

back bias control. However, there are no previous studies on applying RBB to logic-compatible GC-eDRAMs, which is the subject of Sect. 5.3. Silicon measurements of a 2 kb GC test array implemented in 0.18 μm CMOS technology show an improvement of the retention time by up to two orders-of-magnitude when sweeping the body voltage of the transistor over a range of 375 mV.

The third and final technique, presented in Sect. 5.4, is using replica GCs to track the data integrity of the actual GC array across PVT corners and across varying write access statistics to account for write disturbs. Refresh timing is conventionally set according to the worst-case PVT corner and write disturbs, leading to unnecessarily high refresh rates under most operating conditions. This chapter introduces a replica GC technique which tracks the actual PVT and write disturb conditions and triggers an array refresh only when needed, thereby reducing the data retention power under most operating conditions. Silicon measurements of a 2 kb GC test array with integrated replica GCs, implemented in 0.18 μm CMOS technology, show efficient retention time tracking and refresh rates, which are over 5× lower than conventional worst-case refresh triggering.

5.2 Single-Supply Transmission-Gate (TG) 3T-Bitcell GC-eDRAM

Generally speaking, the data retention time (DRT) of GC-eDRAMs is the time interval from writing a data level into the bitcell to the last moment at which one can still correctly read out the stored information. Note that more detailed mathematical definitions of DRT were provided in Chaps. 3 and 4. The DRT is primarily limited by the initial charge stored on the storage node capacitor (C_{SN}) and the leakage currents that degrade the stored voltage level over time. For traditional 2T and 3T GCs, the DRT is significantly affected by the initially degraded voltage level corresponding to data "0" or "1" due to the threshold voltage (V_T) drop across the write transistor (MW in Fig. 5.1a). In order to address this problem, a boosted write word-line (WWL) voltage is usually employed to pass a full-swing data level to the SN. However, this requires the generation of a boosted on-chip voltage, which entails a substantial overhead [10]. The magnitude of the voltage boost is set not only to overcome the V_T drop, but also to achieve short write-access times, which otherwise are typically longer than for 6T SRAM. Furthermore, charge injection (CI) and clock feed-through (CF) during WWL signal de-assertion cause a voltage step at the SN, resulting in an initially degraded level at the end of a write access [9]. This undesired voltage step increases with WWL boost magnitude, which results in a trade-off between write speed and DRT (refresh power) [9]. In addition, the level-shifting and tolerance of higher than nominal voltages can be complex, especially when this boosted voltage is a negative underdrive voltage, as required by GCs using a PMOS write transistor [6, 11]. The propagation of such a negative voltage can easily lead to voltage drops over device terminals that violate the technology limitations.

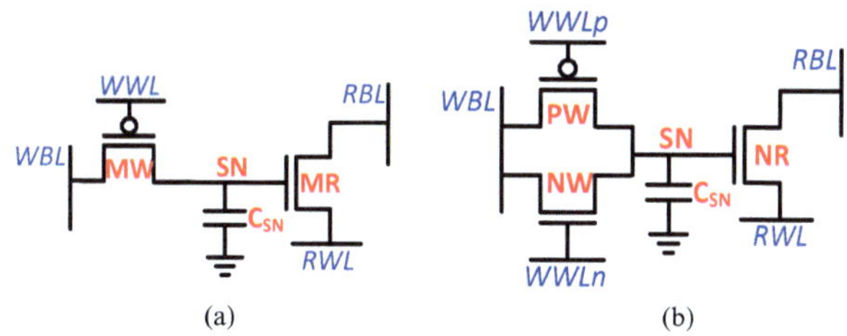

Fig. 5.1 Schematic representations of conventional and proposed GCs. (**a**) 2T Mixed PMOS/NMOS GC. (**b**) Proposed 3T TG GC

In this section, a new topology for a 3T GC is introduced, featuring a complementary NMOS plus PMOS transmission gate (TG) in the write port. The proposed 3T GC provides strong initial data levels (both "1" and "0") for enhanced DRT and robust operation, as well as fast write-access times. These advantages are achieved without the need for additional voltages or boosted signals, allowing the use of standard peripheral circuits for simple SoC integration and small silicon area.

5.2.1 Proposed 3T TG Gain-Cell

5.2.1.1 3T TG Gain-Cell Structure

Figure 5.1b shows the schematic representation of the proposed single-supply 3T GC. The circuit comprises a write port featuring a complementary TG composed of NMOS and PMOS write transistors (NW and PW), a read port based on an NMOS read transistor (NR), and a SN composed of the parasitic capacitance of the three devices and the stacked metal interconnect. The cell is built exclusively from standard V_T transistors and, like conventional GCs, is fully compatible with standard digital CMOS technologies. The gates of NW and PW are connected to complementary write word-lines, WWLn and WWLp. A common write bit-line (WBL) is used to drive data through the TG during write operations. The capability to pass a full-swing through the TG enables the propagation of strong levels to the SN without the need for a boosted WWL. Read is performed by pre-charging the read bit-line (RBL) and subsequently driving the read word-line (RWL) to ground (V_{SS}), thereby conditionally discharging the RBL capacitance if the SN is high (data "1") or blocking the discharge path if the SN is low (data "0"). To achieve a good trade-off between speed, area, power, and reliability, a dynamic sense inverter is used on the readout path, as expatiated on in Sect. 5.2.2.1. However, other sense amplifiers can be used for improved read performance, such as the ones demonstrated in [1–3].

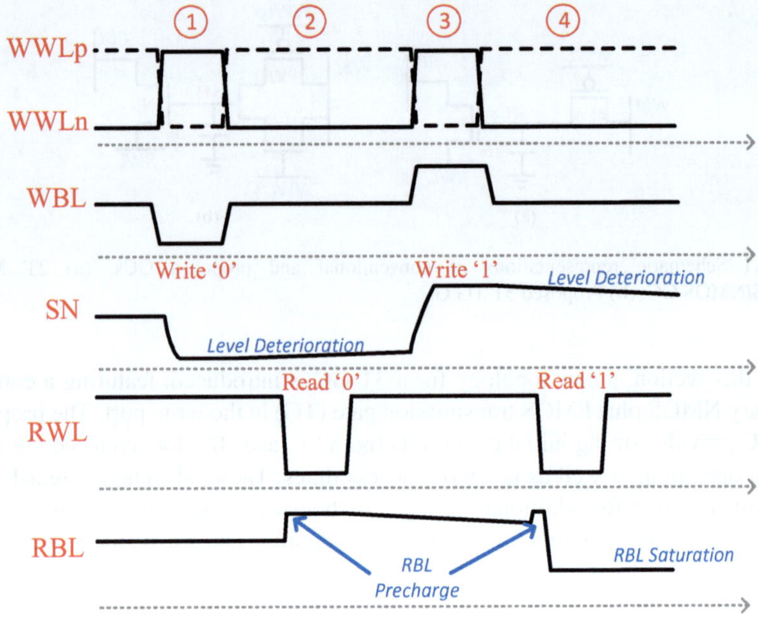

Fig. 5.2 Timing diagram of subsequent write and read operations: ① Write "0"; ② Read the stored "0"; ③ Write "1"; and ④ Read the stored "1." Plots extracted from Spectre simulations with nominal parameter values

5.2.1.2 3T TG Gain-Cell Operation

Figure 5.2 demonstrates bitcell operation through the application of subsequent write and read operations of both data values with V_{DD}= 900 mV. This supply voltage was chosen as a good median voltage between V_{DD} and V_T, as previously shown to be DRT efficient in GC-eDRAM design [9]. Starting with a charged C_{SN} (①), WBL is driven low and the write word-lines are asserted (WWLp = 0 and WWLn = V_{DD}). As expected, a strong "0" level is passed to the SN, and this level is retained with the de-assertion of the write word-lines due to the opposing CI/CF effects of the PW and NW transistors. During standby, the SN level deteriorates due to leakage currents, dominated by the subthreshold (sub-V_T) leakage of NW and PW in mature CMOS nodes. Therefore, in order to extend the retention time, WBL is driven to $V_{DD}/2$ during standby and read cycles, thereby significantly reducing the sub-V_T leakage through the TG, for both stored data "0" and "1," compared to the case where WBL is driven to either V_{DD} or V_{SS}. Note that Chap. 4 describes the benefit of driving WBL to a voltage level between V_{DD} and V_{SS} for a conventional 2T GC in detail. The WBL biasing circuitry for the 3T TG GC is described in Sect. 5.2.2.1.

During readout (②), the "0" level blocks the discharge path through NR, maintaining the pre-charged voltage on RBL. During the next write operation (③),

WBL is driven high, resulting in a strong "1" stored on the SN. The subsequent read operation (④) provides a strong gate overdrive to transistor NR, thereby discharging RBL in order to read a "1." It should be noted that during this operation (Read "1"), GCs storing "1" and sharing the same column turn on when RBL discharges by more than the V_T of NR, causing the RBL to saturate before it can fully discharge. This phenomenon is common to many GC-eDRAM configurations, as discussed in [9].

5.2.1.3 Comparison to Other Gain-Cell Implementations

A major advantage of the proposed GC over previous 2T and 3T GCs is the self-dampening effect of CI and CF during write, as demonstrated in Fig. 5.3. For a 2T GC with a PMOS write transistor (Fig. 5.3a), CI and CF cause a significant positive voltage disturb on the SN during the rising edge of WWL, resulting in a

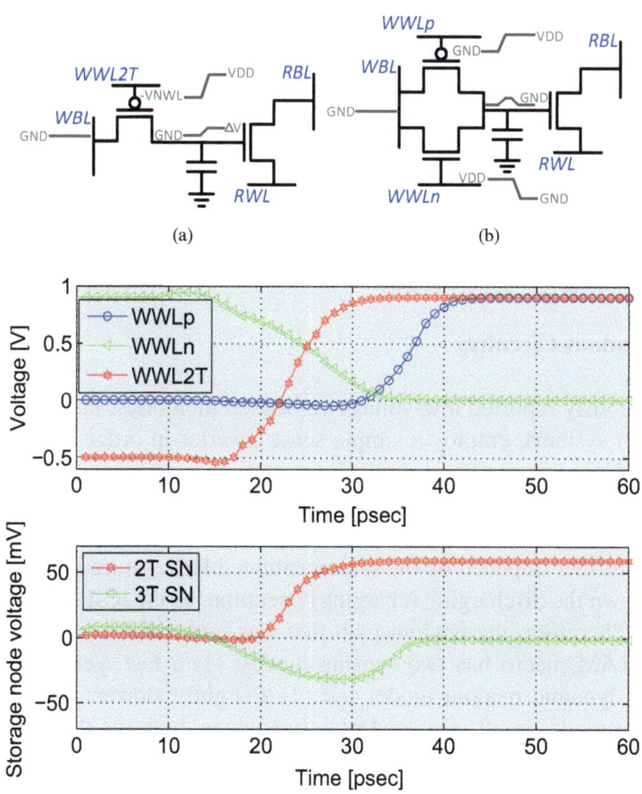

Fig. 5.3 Effects of charge injection and clock feed-through. (**a**) Conventional 2T GC. (**b**) Proposed 3T TG GC. (**c**) Waveform comparison of the two GCs during write de-assertion

degraded initial "0" level. However, in the proposed 3T GC (Fig. 5.3b), this problem is avoided due to the opposite transitions of the complementary write word-lines (WWLn and WWLp), and the opposite polarity of charges injected into C_{SN} from the NMOS and PMOS write transistors. This behavior is shown in Fig. 5.3c for the two GC topologies with V_{DD}= 900 mV and an underdrive voltage of –500 mV for the 2T GC in order to pass a full "0" level on the falling edge of WWL [9]. Whereas the 2T GC suffers from a degraded level of over 50 mV after WWL de-assertion, the CI and CF effects of the complementary devices in the 3T GC essentially negate each other, resulting in a strong 0 V level for this GC. Note that there is still a dip in the 3T GC SN voltage due to the WWLn signal transitioning earlier than the WWLp signal. This could be avoided with more careful timing control. The strong initial data level leads to advantages in terms of both DRT and read-access time.

5.2.2 *Peripheral Circuits*

The single supply voltage required for operation of the proposed 3T TG GC topology simplifies the implementation of a full memory macro and is a significant advantage over other GC-eDRAMs using conventional bitcells, which require level shifters to create the desired boosted or negative (depending on the type of write transistor) WWL voltages. However, to further improve the array performance, in terms of access time and power consumption, several peripheral techniques were integrated into the designed 3T TG GC-eDRAM macro. These peripheral circuits and techniques are presented in the following sections.

5.2.2.1 Readout Circuitry

Several previously reported low-voltage embedded memories, targeted at ultra-low power (ULP) systems, employ a simple sense inverter in order to provide robust, low-area, and low-power data readout. However, such a sense inverter results in slow readout, as it requires the RBL to be discharged (charged for a PMOS read device) past the switching threshold of the inverter, which is often close to $V_{DD}/2$. This operation is further impeded by the aforementioned RBL saturation during readout that slows down the discharging (charging) operation, as the RBL voltage decreases (increases). Therefore, the readout path that was integrated into the proposed 3T TG GC-eDRAM macro has two sensing modes: (1) a fast, yet potentially more error-prone, dynamic readout mode; and (2) a slightly slower, yet more reliable, static readout mode. In all measured prototype chips, both the dynamic (preferred for high-speed operation) and static modes were tested successfully. For both the dynamic and static sensing circuits, low threshold-voltage (LVT) PMOS transistors were used in order to tune the process options for fast read access time. The supply voltage to the readout circuits is gated with the read enable signal in order to save substantial static power due to the leaky LVT transistors. The schematics of the two

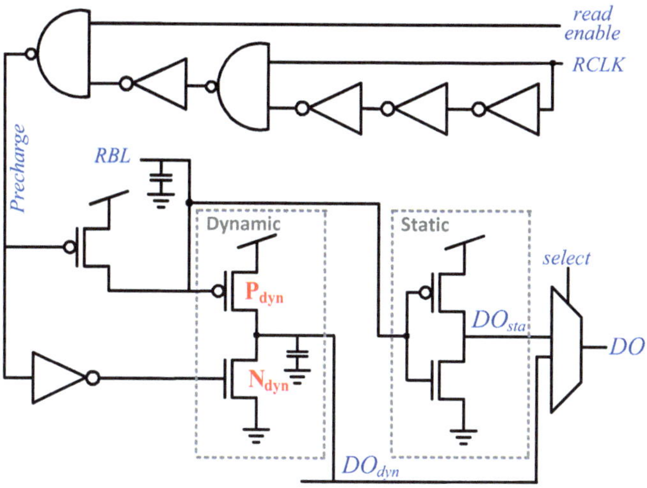

Fig. 5.4 Dynamic and static readout circuits

described readout circuits are shown in Fig. 5.4. The rising edge of the read clock (RCLK) creates a pre-charge pulse, which charges the parasitic capacitance of RBL and discharges the output capacitance of the dynamic sense inverter (DO_{dyn}) through the discharge transistor, N_{dyn}. Subsequently, RBL is conditionally discharged during the read operation, turning on P_{dyn} to flip the output if a "1" is stored in the selected GC. Therefore, an RBL swing of only one V_T is required to complete a read operation. Transistor sizes for the dynamic sense inverter and the pulse generator were chosen according to post-layout simulations, accounting for global and local parametric variations.

5.2.2.2 Write Circuitry

The proposed single-supply 3T TG GC provides a significant improvement in both write time and initial SN level compared to standard GC implementations. However, the dual-transistor write port adds an additional leakage path to/from the SN. The increased aggregated sub-V_T current causes faster degradation of the stored charge, leading to reduced DRT, as compared to a reference 2T GC. In addition, as discussed in Chap. 4, the asymmetric DRT for data "0" and "1" of a standard 2T GC can be manipulated to enhance the DRT by biasing the WBL at the best-case voltage for the weaker data level during standby and read operations [15]. For the proposed 3T TG GC, the worst-case DRTs for data "1" and "0" are similar, and significant deterioration of the stored data levels occurs for both extreme values of WBL bias (V_{DD} and V_{SS}). However, a neutral bias voltage of $V_{DD}/2$ can be applied to the RBL during standby and read operations, thereby greatly reducing the aggregated sub-V_T current through the TG. Figure 5.5 illustrates the benefit of this WBL

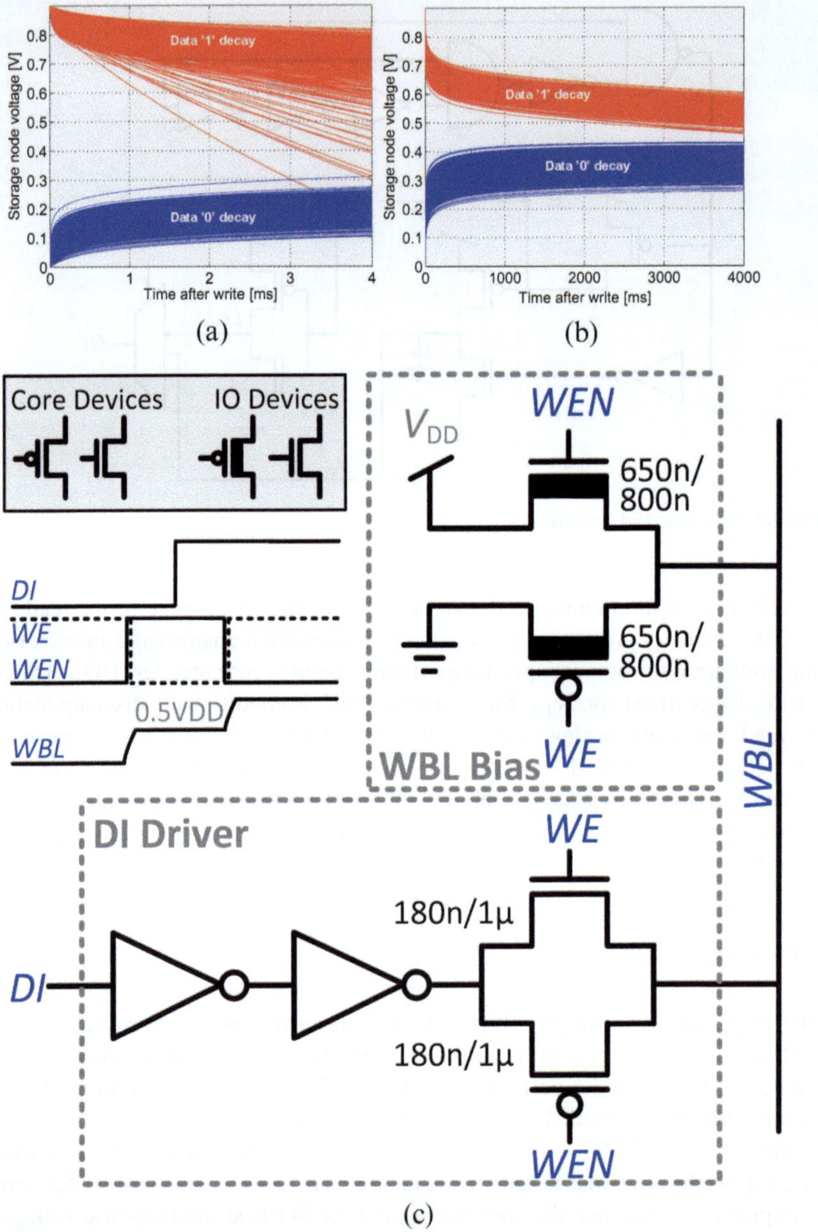

Fig. 5.5 Improved DRT achieved by $V_{DD}/2$ WBL biasing and the corresponding biasing circuit. Plots include 1k Monte Carlo samples at 27 °C and V_{DD} = 0.9 V. (**a**) SN degradation with worst-case biasing conditions. (**b**) SN degradation with $V_{DD}/2$ WBL biasing. (**c**) Write circuitry for $V_{DD}/2$ bias generation

bias, by showing the level degradation of data "1" and "0" with WBL biases of opposite polarity and $V_{DD}/2$, according to 1k Monte Carlo samples simulated at 27°C and $V_{DD} = 0.9$ V. With a WBL bias of $V_{DD}/2$, the DRT can be extended by approximately three orders-of-magnitude.

The write circuitry to generate the $V_{DD}/2$ WBL bias during standby and read cycles is shown in Fig. 5.5c. A standard inverter chain conditionally drives the data-in (DI) level onto the WBL through a TG, controlled by complementary write enable signals (WE and WEN). In parallel, a pair of long-channel I/O devices drive WBL during non-write cycles. These devices create a short-circuit path between V_{DD} and V_{SS} when WE is low, providing a $V_{DD}/2$ voltage at the WBL. Due to the thick oxides and long channel lengths of the I/O devices, process variations are significantly reduced, and the static current consumption is extremely low. Based on the chosen transistor sizes, the static power consumption of the proposed WBL driver during non-write cycles is only a few fW for a 900 mV supply voltage, which is negligible compared to the refresh power of the GC-eDRAM array.

5.2.3 Macrocell and Test Chip Design

A 64×32 bit (2 kb) memory macro based on the proposed 3T TG GC was designed in a standard 0.18 μm CMOS process and integrated into a larger test chip with various test structures. All devices were implemented with standard core or I/O transistors to provide complete logic process compatibility without the need for additional process steps. Minimum-sized write transistors were used in order to achieve a small bitcell area, while a slightly wider read transistor was employed in order to improve the read-access time. The WWLs were routed with horizontal polysilicon stripes in order to provide a dense bitcell. All other signals were routed in the lowest three metal layers. Higher interconnect layers (metals 4 and 5) were tightly stacked above the bitcell in order to increase C_{SN}. The resulting bitcell, shown in Fig. 5.6a, has an area of 6.97 μm^2 (1.49 μm × 4.68 μm). This is 43% smaller than a 6T SRAM bitcell drawn according to the same design rules, shown in Fig. 5.6b.

A micrograph of the fabricated test chip is shown in Fig. 5.6c. In addition to the 2 kb 3T TG GC-eDRAM array, the test chip includes a built-in self-test (BIST) for at-speed functionality tests, a 2 kb SRAM macro for storing per-bit data comparison results with expected responses, and several other test structures. The test chip was designed to enable three primary test modes: (1) full at-speed testing using the BIST; (2) single-operation array control through scan chain configuration; and (3) partial direct array access through external signals. These three test configurations were used to test the functionality of the 3T TG GC-eDRAM array and to extract the measurement data shown in the next section. The GC-eDRAM array was biased by a separate low supply voltage (MV_{DD}), while the supply voltage of the digital core (V_{DD}) was kept higher to ensure functionality and high speed of the BIST and digital control circuits.

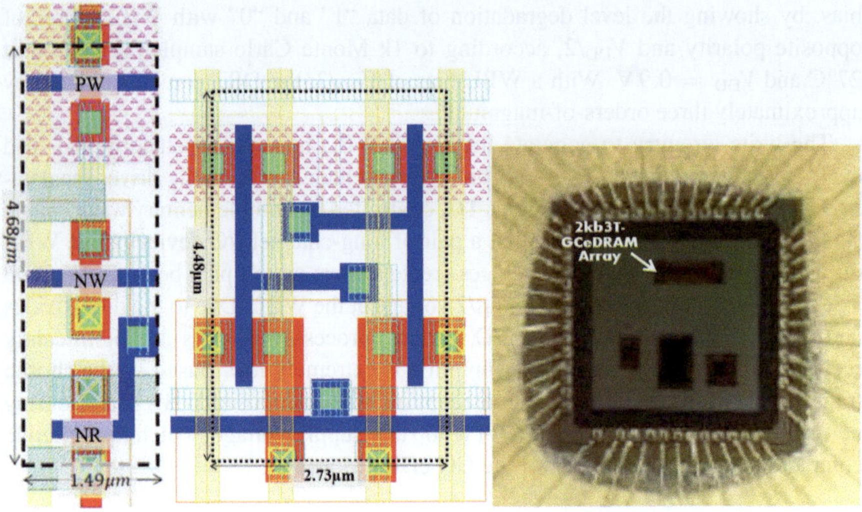

(a) (b) (c)

Fig. 5.6 (**a**) 3T gain-cell layout. (**b**) Redrawn 6T SRAM layout. (**c**) Test-chip micrograph

The primary test starts by serially loading configuration registers and is followed by launching the BIST to enable at-speed functionality tests of the proposed 3T TG GC-eDRAM with various data values, refresh rates, and operating frequencies. The general testing procedure of the BIST includes writing a predetermined data sequence into the GC-eDRAM array and then moving into an idle state for a configurable duration. In the idle state, special *write disturb* cycles can be activated to emulate the worst-case DRT. This is achieved by driving the WBLs to a voltage level opposite to the stored data, maximizing the sub-V_T leakage from/to the SN, emulating worst-case write access to different GCs in the same column. The configurable percentage of write disturb cycles provides a means to measure the DRT for different GC-eDRAM access statistics, rather than just assuming the nonrealistic worst-case write access during 100% of the time. Following the retention period with configurable write disturb, the array data is read out, compared with the initially written data, and stored in the on-die 2 kb SRAM macro. This allows the extraction of detailed per-cell GC-eDRAM DRT, write failure, or read failure maps under marginal conditions. The other test modes enable fine-grained debugging, as well as flexible application of test sequences not planned for at design time.

5.2.4 Lab Setup and Silicon Measurements

The test chips were packaged and measured using the test procedures described above. All ten packaged chips were fully operational across the complete range of supply voltages (MV_{DD}) from 600 mV to 1.8 V. In order to study per-bit DRT, a single value was written to the entire array and the array was then put into standby for a given time period (t_{ret}). Following this retention period, the array was read out and compared to the written data values. This procedure was repeated for several t_{ret} values and the measured bit DRT was calculated as the first t_{ret} that caused a read failure for each particular bit. In order to better emulate a typical operating scenario, a 5% write disturb was applied to the array. While the DRTs were measured for both all-"0" and all-"1" data levels, only all-"1" measurements are shown, since it was found to be the worst case of the two. This corresponds with Fig. 5.5, showing that data "1" degrades faster than data "0," while also requiring a longer sensing period for the chosen read circuitry. The resulting retention map for one of the measured chips is shown in Fig. 5.7a. All measurements were taken at room temperature, which is considered a typical temperature for ULP applications that do not suffer from self-heating due to low computational complexity. DRT is shown on a $\log_{10}$ scale to better visualize the difference between cells, as it varies over several orders-of-magnitude. The lack of a systematic pattern shows that the difference between cells is primarily due to local process variations. The wide distribution of DRT is shown in Fig. 5.7b for all bitcells of ten measured dies (a total

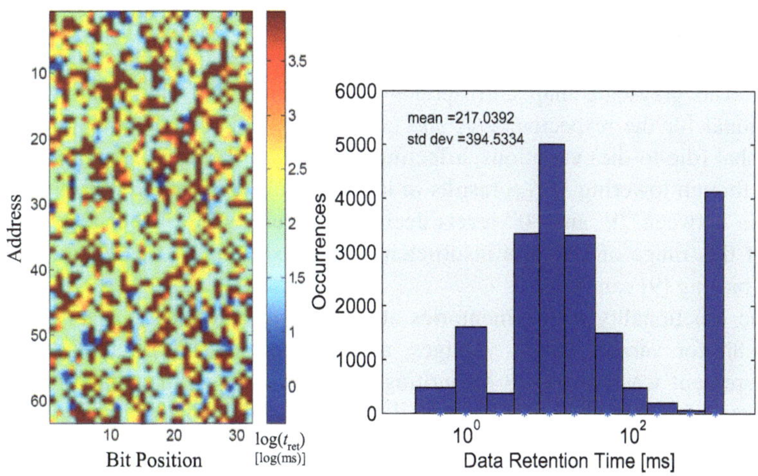

(a) (b)

Fig. 5.7 (**a**) Retention time map of a 2 kb 3T GC array with MV_{DD} = 900 mV. (**b**) DRT distribution of 20,480 3T GCs

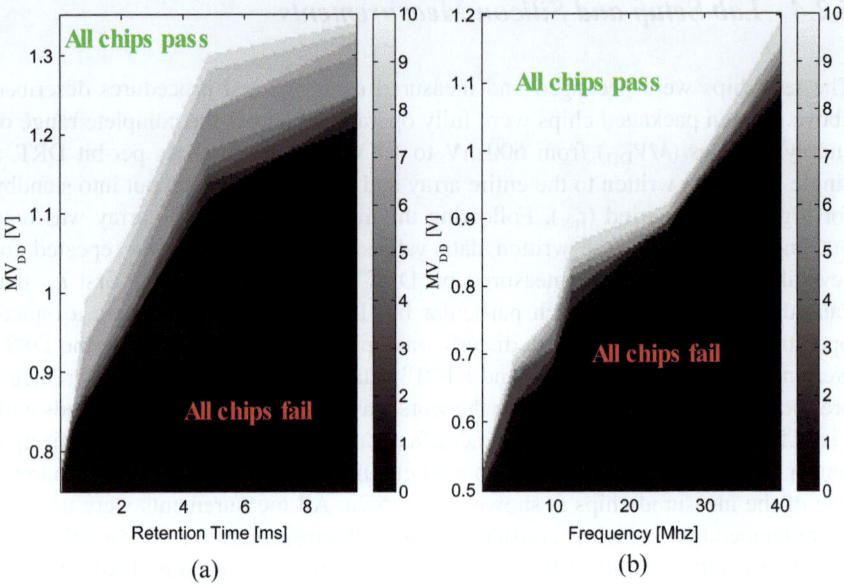

Fig. 5.8 Minimum memory supply voltage Shmoo plots for ten measured chips. (**a**) Min. MV_{DD} vs. DRT. (**b**) Min. MV_{DD} vs. frequency

of 20,480 cells). The minimum and maximum DRTs were found to be 0.8 ms and 1978 ms, respectively. The large spread and lack of systematic pattern correspond with previous studies [9, 16].

Figure 5.8a shows a Shmoo plot of MV_{DD} vs. the measured DRT for ten different chips. The grayscale map corresponds to the number of chips that were fully functional for the respective V_{DD} and targeted DRT. The gray levels are a result of global (die-to-die) variations, affecting the DRT of the array for a given MV_{DD}. Even though lowering MV_{DD} results in lower leakage currents from/to the SN, the margin between "1" and "0" levels decreases, resulting in a lower DRT. We note that if this range of DRTs is insufficient, previously reported techniques such as body biasing [9] can be used.

The functionality of the memories at different frequencies is demonstrated in Fig. 5.8b for various MV_{DD} voltages, as measured for 10 different chips. The measurement was conducted by writing "1" to all bits of the array and reading them out after the previously measured maximum DRT. The operating frequency was swept for every MV_{DD} to provide the Shmoo plot. The array showed full functionality for all frequencies up to 40 MHz with a supply voltage of 1.2 V. Measurements at higher frequencies were impossible due to limitations of the test setup, which was designed for low frequency ULP applications.

The memory frequency is limited by the read-access time, since during readout, RBL needs to discharge in order to flip the sense inverter. This operation strongly depends on the parasitic capacitance of the RBL, as well as the data held in the unselected cells sharing the same column. If any of these cells stores a "1," the RBL discharge saturates at around V_{DD}–V_T, since at that point the unselected NRs start conducting and counteract the discharging efforts of the selected bit. In addition, due to the fact that readout was performed after the maximum DRT for each V_{DD}, a worst-case scenario was measured, since the data in the cells was already deteriorated, making the drive current much lower compared to the case where strong data levels are stored in the cells. Raising V_{DD} clearly improves the read-access time due to a stronger drive current, which makes the sense inverters switch faster.

For a dynamic memory, the relevant metric for static power consumption is retention power, composed of the sum of leakage and refresh power. Retention power was measured for storing an entire array of "0" and "1" under supply voltages ranging from 600 mV to 1.2 V. Leakage power mainly consists of sub-V_T currents from the SN to the WBL, which is biased at $MV_{DD}/2$ during standby periods. The measured retention power was found to be 4.9 nW for MV_{DD}= 900 mV and 1.01 μW for MV_{DD}= 1.2 V, which is 17× lower than a previously reported 6T SRAM cell [12], operated at 1.2V in the same technology node (consuming 26 μW standby power with V_{DD}= 1.8V). The refresh frequency for every MV_{DD} was selected according to previously measured minimum DRTs.

5.3 Impact of Body Biasing (BB) on Retention Time

Reverse body biasing (RBB) is a well-known technique to suppress leakage current and is extensively and industrially used in conventional 1T-1C DRAM technology. In fact, the p-well is biased to a negative voltage to improve the data retention time in most DRAM chips, a technique also referred to as back bias control. However, there are no previous studies on applying RBB to logic-compatible GC-eDRAM in order to improve its retention time and reduce its data retention power. In the following, we measure the impact of body biasing as a control factor to improve the retention time of a 2 kb GC-eDRAM macrocell, and also examine the distribution of the retention time across the entire GC-eDRAM array. The concept is demonstrated through silicon measurements of a test chip manufactured in a standard CMOS 0.18 μm process. While there is a large retention time spread across the measured 2 kb gain-cell array, the minimum, average, and maximum retention times are all improved by up to 2 orders-of-magnitude when sweeping the body voltage over a range of 375 mV.

As already mentioned in Sect. 2.3, the main drawback of GC-eDRAMs is the need for periodic refresh, since refreshing the memory results in a considerable amount of power consumption and limits the read/write availability of the memory array. Therefore, to improve the competitiveness of GC-eDRAM, it is crucial to extend the data retention time. Data levels in GC-eDRAMs are stored as charge on the capacitive storage node (SN), whose equivalent capacitance is referred to as C_{SN}, and therefore data retention is limited by the time it takes for this charge to leak away. Several simple measures can be taken to extend the retention time, such as:

1. Increasing C_{SN} through layout techniques. These techniques include increasing the diffusion area of the write transistor, increasing the gate area of the storage transistor, and increasing the coupling capacitance of the metal stack and vias readily available in digital CMOS technologies [7, 9].
2. Minimizing the subthreshold conduction through the write access transistor (MW) by using low-leakage MOS transistors [6].
3. Employing write bit-line (WBL) control schemes to minimize charge loss through MW (see previous Sect. 4.3 and [5]).

An additional technique that was not previously applied to gain-cells is threshold voltage (V_T) adjustment through body biasing. The application of a reverse body bias (RBB) increases V_T, and therefore reduces the charge loss through subthreshold conduction to achieve longer retention times. However, this means of control can also be used to improve the array availability by applying a forward body bias (FBB) during refresh cycles to reduce access time [14]. Our main contributions can be summarized as follows: (1) For the first time, we propose reverse body biasing as a technique to improve the retention time of GC-eDRAM and demonstrate its high effectiveness through silicon measurements; and (2) the retention time penalty of forward body biasing, used for fast memory access and short refresh times, is evaluated.

5.3.1 Bitcell Design for Body Biasing Experiment

Figure 5.9a shows the schematic and the basic operation of the two-transistor (2T) all-PMOS gain-cell used in this study (a similar cell has previously been proposed in [6]). Other than the high-V_T I/O PMOS write transistor (MW) requiring a larger underdrive voltage, the cell operation is equal as for the gain-cell considered in Sect. 4.3 and is therefore recalled only briefly. MW is used to transfer the data driven

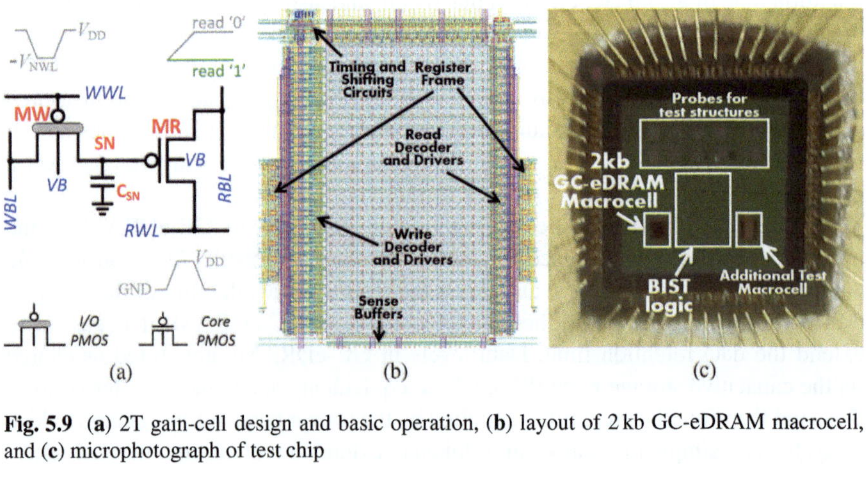

Fig. 5.9 (**a**) 2T gain-cell design and basic operation, (**b**) layout of 2 kb GC-eDRAM macrocell, and (**c**) microphotograph of test chip

onto the WBL to C_{SN}. MR is the read access transistor, used to read out the data level stored in the bitcell. A write access is initiated by applying an underdrive voltage ($-V_{NWL}$) to the write word-line (WWL) in order to properly transfer a logic "0" level (V_{SS}) from WBL to SN in a short time. A read access is initiated by pre-discharging the read bit-line (RBL) and subsequently raising the read word-line (RWL). If a logic "0" is stored on C_{SN}, MR will charge RBL past a detectable threshold. Alternatively, if a logic "1" (V_{DD}) has been written to the SN, RBL will remain discharged below the sensing threshold. The basic C_{SN} is increased by building up side-wall capacitors between the SN and a constant potential (V_{DD}) atop the bitcell footprint, using all available metal layers (a total of six layers in the considered 0.18 μm CMOS process).

The dominant leakage mechanism that causes the deterioration of the stored data levels is clearly the subthreshold conduction of MW. This is especially true for mature CMOS nodes, such as the 0.18 μm process used in this study, but also holds for a deeply scaled 40 nm CMOS node [9] (as will be seen in Chap. 6, focusing on aggressive voltage and technology scaling). In order to achieve the longest possible retention time, an I/O PMOS transistor is used to implement MW, as this device features the lowest subthreshold conduction among all devices offered in the chosen 0.18 μm CMOS technology [7]. By also using a PMOS device to implement MR, the entire array resides in an equi-potential n-well, enabling simple control over the body voltage (V_B) of the bitcells. Reverse biasing the n-well at a voltage above V_{DD} increases the V_T of the transistors, thereby suppressing the subthreshold conduction of MW and improving the retention time. Likewise, forward biasing V_B below V_{DD} lowers the V_T of the transistors, resulting in faster read and write access times. Such forward biasing, of course, also results in larger subthreshold conduction, and therefore decreased retention times.

In this section, the variable ΔV_B is used to express the amount of body biasing, according to $V_B = V_{DD} + \Delta V_B$, where a positive and a negative value of ΔV_B correspond to RBB and FBB, respectively. In this study, a biasing range of $-250\,\text{mV} < \Delta V_B < 125\,\text{mV}$ is considered, corresponding to a V_T range of $-770\,\text{mV} < V_T < -625\,\text{mV}$ for a PMOS I/O device under otherwise nominal conditions with $V_{DD} = 750\,\text{mV}$ (corresponding to a near-V_T supply voltage for core transistors).

5.3.2 *Macrocell Architecture and Test Chip Design*

Figure 5.9b shows the layout of the 2 kb gain-cell macro memory in 0.18 μm CMOS technology. The core bitcell array consists of 64 × 32 gain-cells, all sharing the same n-well and V_B. N-well contacts are provided every 16 rows, as well as at the top and the bottom of the array. In addition to the bitcell array, the macrocell comprises the following peripheral circuits:

1. Write address pre- and post-decoders that drive V_{DD} or $-V_{NWL}$ onto WWL
2. Read address pre- and post-decoders that drive V_{DD} or V_{SS} onto RWL
3. Level-shifters

4. WBL drivers
5. Readout sense buffers
6. Timing control units

Figure 5.9c shows a microphotograph of the manufactured test chip. In addition to the 2 kb GC-eDRAM macrocell in the lower-left corner, the chip contains a built-in self test (BIST) unit. The main features of the BIST can be summarized as follows: (1) address sequence generation (increasing, decreasing, pseudo-random); (2) data pattern generation (checkerboard, pseudo-random, all-"1," all-"0"); (3) programmable refresh period of the memory under test (MUT); (4) pass/fail decision during readout of the MUT; (5) embedded SRAM for storing maps of MUT retention time, read failures, or write failures; and (6) support for two-port operation of the MUT. Finally, the test chip also contains scan chains for full access to the MUT with any data or address sequence pattern independent of the BIST.

5.3.3 Silicon Measurements

The packaged test chips were mounted on a test board by means of a burn-in socket and connected to a TMPC PG3A pattern generator and a Tektronix TLA6403 logic analyzer. The main supply of the memory macrocell was set to 750 mV, and the body voltage V_B was swept from 500 to 875 mV to analyze the impact of body biasing. A separate negative voltage of −1.5 V was supplied to the macrocell for the WWL underdrive. The BIST and other digital control units were supplied with the nominal voltage of the technology, which is 1.8 V. Both the write and read access times were set to 1 μs for robust write and read operations, even at the low V_{DD} of 750 mV. This ensured that the measured failures relate to retention time, and were not caused by incomplete writes or erroneous reads due to insufficient access time. Table 5.1 summarizes the primary specifications of the measurement setup.

Measurements indicate that the 2-PMOS gain-cell retains logic "1" levels for extensive periods (>1 s), even when the WBL is held at 0 V (which maximizes the subthreshold conduction of MW). This coincides with previous reports that logic "1" levels decay very slowly due to the increasing reverse gate overdrive and body effect of MW as the SN voltage drops [7]. Therefore, the retention time of the gain cell is almost exclusively limited by its ability to hold a logic "0" level.

Table 5.1 Measurement setup for GC-eDRAM test chip with adaptive body bias control

V_{DD}	750 mV
ΔV_B	−250 to 125 mV
Write access time	1 μs
Read access time	1 μs
Write-"1" disturb activity	25%
Temperature	Room temperature (uncontrolled)

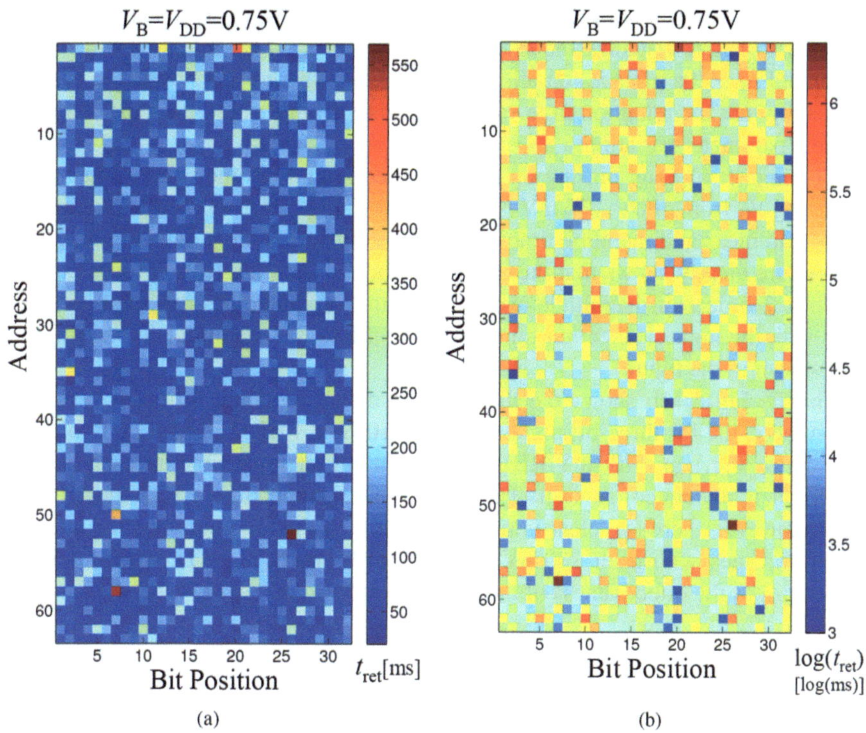

Fig. 5.10 (**a**) Retention time (t_{ret}) map of 2 kb 2T gain-cell array with standard body bias and $\alpha_{disturb} = 25\%$ at room temperature, and (**b**) map of log(t_{ret})

The decay of a cell's logic "0" level is heavily dependent on the state of the WBL. On the one hand, when WBL is low, subthreshold conduction through MW discharges the SN, reinforcing a stored logic "0" level. On the other hand, when WBL is high, a worst-case condition occurs, as leakage through MW causes accelerated decay of a stored logic "0" level. Our measurement setup assumes a 50% write duty cycle (i.e., there is a write access during 50% of the time) and that the probability of writing a "1" (which requires pulling WBL up to V_{DD}) is 50% as well. Overall, this leads to a write-"1" disturb activity factor ($\alpha_{disturb}$) of 25%.

Using the measurement setup described above, retention time was measured for the entire 2 kb array under standard biasing conditions (i.e., $V_B = V_{DD} =$ 750 mV) at room temperature (temperature was not controlled). The results of this measurement are shown in Fig. 5.10a. The minimum and maximum retention times (t_{ret}) of 2048 measured gain-cells were found to be 23 and 569 ms, respectively, corresponding to a ratio of 25 between the maximum and minimum value. A recent study [3] reports an even higher ratio of over 50 between the maximum and minimum measured retention times in an 1 kb array implemented in 65 nm CMOS. In the present study, the majority of the cells exhibited retention times in the range of 20–200 ms (dark and light blue color), whereas a small number of cells exhibited considerably higher retention times (yellow, orange, and red colors). In order to better visualize the differences among the lower retention times (20–200 ms), Fig. 5.10a plots t_{ret} on a logarithmic scale. There is no systematic pattern, indicating that the retention time variability arises from local (within-die), random process parameter variations.

The impact of body biasing on the measured retention times was evaluated by sweeping V_B from 500 to 875 mV (-250 mV< ΔV_B <125 mV). The minimum and maximum measured retention times across the entire array are plotted in Fig. 5.11a. This figure clearly shows that the minimum and maximum retention times change by up to 2 orders-of-magnitude over this 375 mV V_B range. As expected, the best cells with the highest retention time remain at the same location under varying V_B (not shown in the figure).

Finally, Fig. 5.11b shows the distributions of the retention time across the 2k measured cells, for three biasing conditions: 100 mV FBB, standard body biasing (i.e., $V_B = V_{DD}$), and 100 mV RBB. The minimum retention time for each biasing condition is annotated, as well. The spread of retention time across the array is large; however, there is a clear improvement in both the minimum and the average retention times with each 100 mV increase in the body bias. This measurement demonstrates the effectiveness of the proposed technique.

5.4 Replica Technique for Optimum Refresh Timing

The primary component of power consumption in GC-eDRAMs is the dynamic power consumed during periodic refresh operations. Refresh timing is traditionally set according to a worst-case evaluation of the retention time, under extreme environmental variations, namely process-voltage-temperature (PVT) variations, and worst-case access statistics, leading to frequent, power-hungry refresh cycles. In this section, we present a replica technique for automatically tracking the retention time of a GC-eDRAM macrocell according to both PVT variations and operating statistics (access patterns), in order to reduce the data retention power of the array. A 2 kb array was designed and fabricated in a mature 0.18 μm CMOS process, appropriate for integration in ultra-low power applications such as biomedical sensors. Silicon measurements show efficient retention time tracking across a range of supply voltages and access statistics, reducing the refresh frequency by more than 5× compared to traditional worst-case design.

5.4.1 Conventional Design for Worst-Case Retention Time

In order to demonstrate the replica technique for optimum refresh timing, we consider the same all-PMOS 2T GC topology as for the adaptive body biasing study presented in Sect. 5.3. This GC topology is shown again in Fig. 5.12, which also

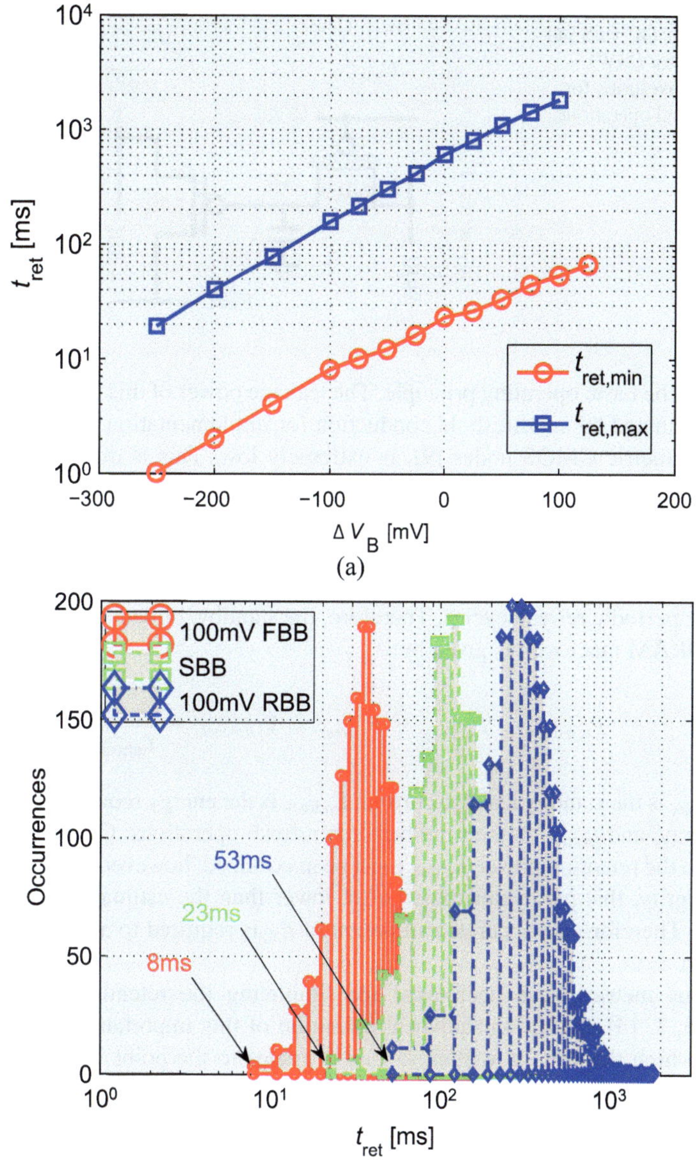

Fig. 5.11 $V_{DD} = 750\,\text{mV}$ with $\alpha_{disturb} = 25\%$ at room temperature: (**a**) Minimum ($t_{ret,min}$) and maximum ($t_{ret,max}$) retention times across the entire 2 kb array, as a function of ΔV_B, and (**b**) retention time distributions of 2048 measured gain-cells for 100 mV FBB, standard body biasing (SBB), and 100 mV RBB

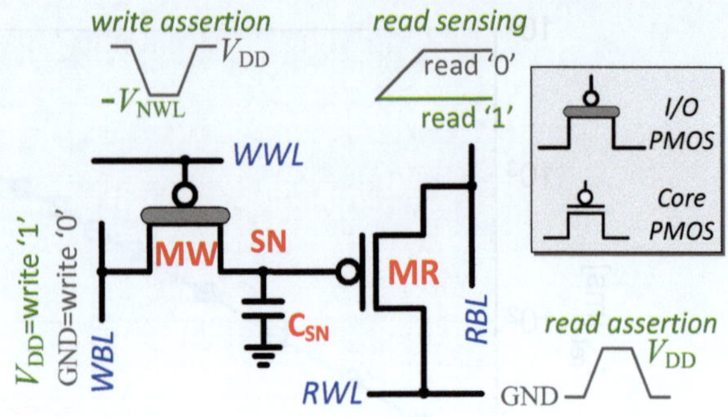

Fig. 5.12 Schematic of the all-PMOS 2T GC with I/O write transistor (MW), including waveforms for write and read operations

illustrates the basic operating principle. The leakage power of this GC circuit, shown to be dominated by subthreshold conduction for implementation in submicron and even nanometric CMOS nodes [9], is extremely low. This is due to the fact that during standby and write, the drain-to-source voltage (V_{DS}) of the read transistor MR is zero, and the subthreshold leakage through the write transistor MW is limited to (dis)charging the storage node capacitor C_{SN}. The obvious issue is that any leakage to or from the storage node SN results in a degradation of the stored data level, requiring periodic refresh cycles. Therefore, the standby, or data retention power of a GC-eDRAM macrocell is given by:

$$P_{\text{retention}} = P_{\text{leakage}} + P_{\text{refresh}} = V_{DD} I_{\text{leak}} + \frac{E_{\text{refresh}}}{t_{\text{refresh}}} \tag{5.1}$$

where I_{leak} is the standby leakage current, E_{refresh} is the energy required to refresh the entire array, and t_{refresh} is the time between refresh operations. Clearly, in order to minimize the retention power, t_{refresh} must be maximized; however, in order to ensure data integrity, this parameter must be set lower than the estimated data retention time t_{ret}. Therefore, an accurate estimation of t_{ret} is required to achieve low power operation.

Various metrics have been used for simulating the retention time t_{ret} of a bitcell [6, 7, 13], but the unequivocal definition of this important parameter is the time at which the voltage written to C_{SN} degrades to the point where it results in an incorrect readout. This time is set by four primary factors: (1) the initial level stored on C_{SN} following a write; (2) the size of C_{SN}; (3) the leakage currents to and from SN; and (4) the readout mechanism. All of these factors are significantly affected by both environmental and manufacturing variations, as demonstrated by silicon measurements in [6]. This results in a large spread of the per-cell retention time across the GC-eDRAM array [3, 8] (see Fig. 5.10a in Sect. 5.3.3). As with any memory array, this necessitates design for the worst cell. However, in addition to the effects of PVT variations, SN leakage currents are highly sensitive to the biasing level of WBL. For a stored "1," the highest discharge leakage occurs when WBL is low, while the worst case for a stored "0" occurs when WBL is high. As shown in [6, 9, 13], the worst-case biasing for a stored "0" leads to a much lower retention time than that for a “1” in an all-PMOS 2T cell. Consequently, in order to determine $t_{refresh}$, the retention time needs to be calculated assuming that WBL is constantly held high. However, this situation would only occur if a write “1” operation was executed on a given column during every clock cycle, leading to early, power-consuming refresh operations in any typical scenario.

5.4.2 Replica Technique Concept

The design for worst-case conditions, coupled with the wide spread of t_{ret} due to PVT variations and write access disturbs, almost always results in the initiation of refresh cycles when the stored data is still at strong levels. By implementing a replica technique to track the global parametric variations, changes in the supply voltage, environmental conditions (such as the temperature), and acute operating characteristics (specifically write accesses), a significant amount of the refresh power can be saved. An additional post-silicon calibration step is implemented to adjust the tracking mechanism for each manufactured die to handle local parametric variations.

The foundation of the proposed technique relies on the superiority of the retention time for data “1” in the all-PMOS 2T gain-cell. This superiority is due to a number of factors, starting with the PMOS write transistor that easily passes a high level to SN, as opposed to a low level, which requires a WWL underdrive to completely discharge C_{SN} in a reasonable amount of time. Subsequently, both charge injection from MW and the coupling capacitance between WWL and SN drive charge onto C_{SN} during the rising edge of WWL (i.e., during the de-assertion of WWL at the culmination of a write operation). This causes a slight voltage rise on SN, resulting in a degraded initial “0” state and an overcharged initial “1” state. Moreover, the decay of a “1” level due to the subthreshold conduction of MW is self-limited due to the steady increase of the reverse gate overdrive and the increasing body effect of MW with progressing decay. A more detailed description of this self-limiting effect will be provided in Chap. 6.

Two primary mechanisms are incorporated to simultaneously extend the retention time of the entire array while maintaining data stability. First, during all non-write cycles, WBL is driven low, thereby enhancing the level of a stored “0” bit while minimally affecting the level of a stored “1.” Second, several replica cells are integrated within an extra column in the GC-eDRAM array and are periodically read out to analyze the state of the storage array’s data retention. These replica cells are standard all-PMOS 2T bitcells, designed with slightly reduced C_{SN} (less metal stacking above the bitcells) to make them fail before the data cells, while tracking the PVT variations of the fabricated array. In addition, the replica column is designed to track the access statistics of the array, rather than assuming unlikely worst-case conditions (i.e., write operations during every clock cycle). Immediately prior to an array refresh, data “0” is written to all of the replica cells, and during read and standby cycles, the WBL of the replica column is driven low, exactly as the WBLs of all columns in the main storage array. Significant data level degradation only occurs when the WBL is high, which can only happen to a cell storing a “0” when a “1” is written to a cell on the same column. Therefore, during write cycles, the WBL of the replica column is driven high, thereby applying worst-case conditions only when they can actually occur. In this way, the retention time of the replica cells is always slightly worse than for a data cell in a column with cells that were repeatedly written as “1” over the retention period. However, instead of assuming an extreme worst case of tying WBL to “1” (which would only occur if the array was written to during every clock cycle), this setup tracks the actual frequency of write operations. Therefore, the replica cells track the access statistics of the array (i.e., the relationship between non-write and write operations), while still ensuring that the replica cells will fail before the real data is lost.

The mechanisms described above accurately track the global variations and access statistics of the array. However, local variations may result in a worst-case retention time lower than that of the worst-case replica cell. Therefore, a post-silicon calibration is used to skew the retention time of the replica cells below the measured worst-case retention time of the array. This is done by employing periodic *pseudo-write* cycles to the replica column. During these operations, the WBL of the replica column is charged, causing the replica cells to degrade at a slightly higher rate than dictated by the write statistics, thereby ensuring the initiation of an array refresh prior to a data loss in the worst cell of the array.[1]

5.4.3 *Replica Technique Integration into Gain-Cell Array*

The proposed replica technique was integrated into a 2 kb all-PMOS GC-eDRAM array in an 0.18 μm CMOS technology according to the schematic illustration in Fig. 5.13. A total of 32 replica cells were placed in an additional column to deal with the large distribution of local variations [3]. In order to maintain the mirrored-column symmetry of the array, a dummy column was attached to the replica column. All replica cells are written with data “0” upon the assertion of the external *RefreshReplica* signal within a single clock cycle, independent of the operation of the rest of the array. The same write mechanism as used for the data bit WWL drivers is incorporated for driving the negative write voltage to the replica cells. The WBL of the replica column is tied to the write enable (WEN) signal in order to track the write statistics of the array. The layout of the replica cells is

[1]In an extremely unlikely case this calibration would be insufficient. This would happen if a continuous write “1” operation was applied to a column with a bitcell with worse retention time than the worst replica cell. However, this scenario would hardly ever occur in any real application. Otherwise, it is still possible to impose a write access policy to the array, which, for example, allows to write to the array only every second clock cycle. In addition, to avoid such a write access policy, it is possible to limit the WBL pulse time for the storage array, while using a pulse width equal to a full clock cycle for the replica columns.

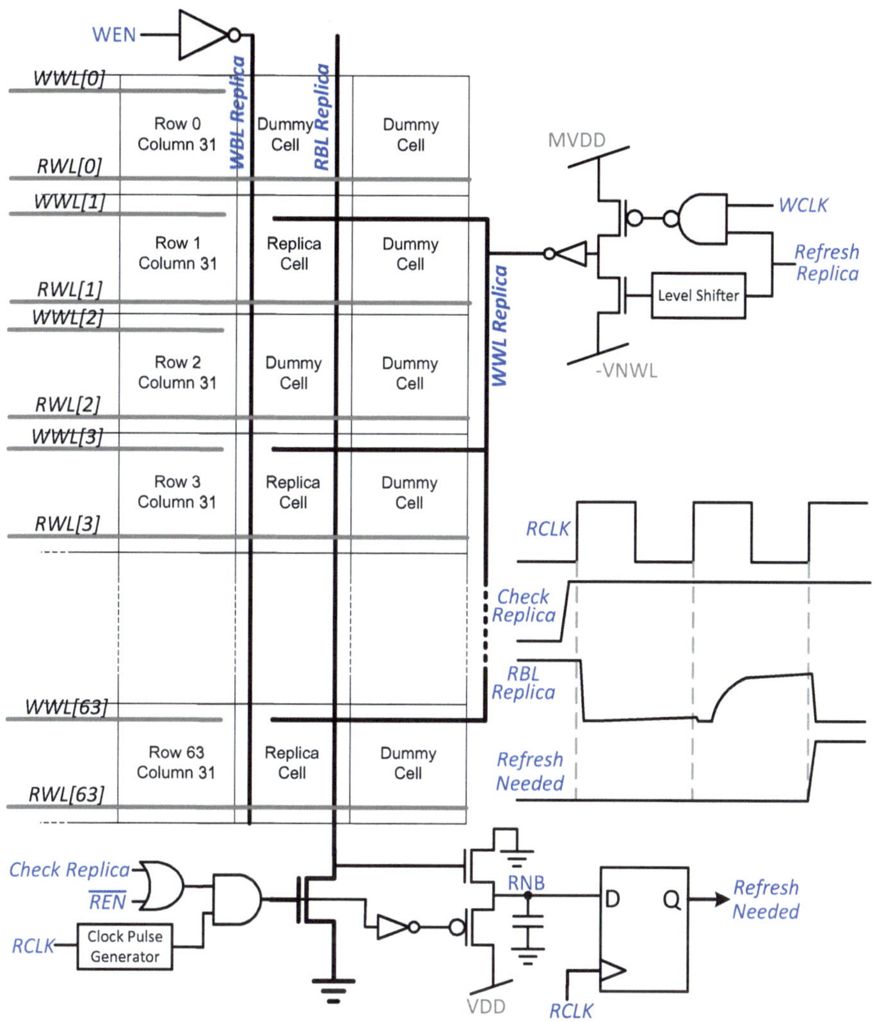

Fig. 5.13 Schematic illustration of the read and write circuitry for operation and control of the proposed replica technique, including timing diagrams

almost identical to the one of standard storage cells; only one layer of the metal stack is removed to reduce the C_{SN} of the replica cells and ensure a slightly lower retention time than for the regular storage cells.

Readout of the replica cells is achieved through a mechanism similar to the readout of the data cells with the addition of a designated *CheckReplica* signal. As the replica cells were designed to fail due to the deterioration of a stored "0" level, reading out a "1" from the replica column indicates the need for a refresh cycle. Therefore, the readout of such an erroneous level is propagated to the control block as the *RefreshNeeded* signal.

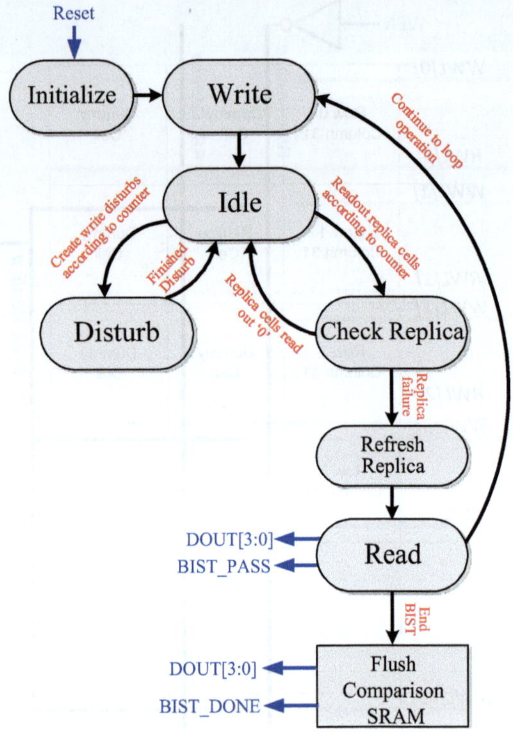

Fig. 5.14 State machine of the test controller

5.4.4 Testing and Characterization Procedure

Testing and characterization of the replica technique were implemented with an on-chip controller, incorporating the finite-state machine (FSM) illustrated in Fig. 5.14. This controller initially writes data to the entire array, and subsequently proceeds into an *Idle* (standby) state for a configurable time period. In the *Idle* state, the controller initiates one of two operations. To measure tracking of write statistics, the controller initiates periodic *Disturb* cycles, during which a row of "1"s (0xFFFFFFFF) is written to a predetermined "victim" address.[2] This *Disturb* operation drives the WBL of all columns high, thereby causing deterioration of stored "0" bits in the entire array. A similar mechanism is incorporated through a post-silicon calibration to further deteriorate the replica cells in order to account for local variations that may otherwise skew the retention time of the worst cell in the array below the retention time of the worst replica cell.

The second operation which can be periodically initiated from within the *Idle* state is the *CheckReplica* sequence, during which the 32 replica cells are serially

[2]The victim address will always store 0xFFFFFFFF, and therefore is not considered for comparison with expected responses.

read out to determine the onset of a refresh operation. If the *RefreshNeeded* signal is asserted (i.e., the data in at least one of the replica cells reads out erroneously), the controller proceeds to refreshing the replica cells, before refreshing the actual storage array and looping back to the *Idle* state for another retention period.

The *Read* state (part of the array refresh sequence) of the test controller provides important measurement data for analysis. The read-out data is compared with the originally written data to ensure equality, and the per-bit comparison results are stored in an on-chip 2 kb SRAM. Concurrently, the one-bit comparison result of the currently read row is driven off-chip via the *BIST_PASS* signal, and the four MSBs are propagated to the external *DOUT*[3:0] pads to enable further observation. An external interrupt signal can break the refresh loop, sending the controller into its termination state, during which the comparison data can be analyzed. In this state, the *BIST_DONE* signal is raised, and subsequently, the full, per-bit comparison data that was stored in the SRAM is flushed out to the *DOUT*[3:0] pads by means of scan chains. This control scheme enables at-speed testing of the GC-eDRAM array, including the ability to observe the functionality of the replica technique under various write disturb statistics.

5.4.5 Silicon Measurements

A 2 kb (64×32) GC-eDRAM array with integrated replica technique was designed and fabricated in a commercial 0.18 μm CMOS technology, as part of the test chip shown in Fig. 5.15. In addition to the array, the test chip included the on-chip test controller, the 2 kb SRAM for data comparison, and several other test components. The test chip was designed to enable three primary test modes: full, at-speed, controller testing; array operation through scan chain configuration; and external direct access to the array. A combination of these three modes was used to test the functionality of the array and produce the measurement data shown below.

The GC-eDRAM bitcell was laid out in a compact array with mirrored rows and columns, as shown in Fig. 5.16, with a unit cell size of 3.024 μm^2 (1.8 $\mu m \times 1.68$ μm). The array, including peripheral circuits, occupies 0.013 mm^2 (106 $\mu m \times 129$ μm) and is biased by a separate, low-voltage supply (MVDD) different than the supply (VDD) of the BIST and the other digital peripheral circuits of the test chip. In addition, an external negative voltage is supplied for write underdrive.

Figure 5.17 illustrates the ability of the replica technique to automatically track the retention time of the array. The figure shows the automatically triggered refresh period for various supply voltages, as compared to the minimum retention time measured at this voltage, following a post-silicon adjustment in the write disturb frequency to account for local variations. Refresh is consistently initiated just prior to the minimum retention time of the array for a range of supply voltages.

Tracking of the write statistics is shown in Fig. 5.18. This figure plots the automatic refresh timing, as compared to the measured retention time of the array

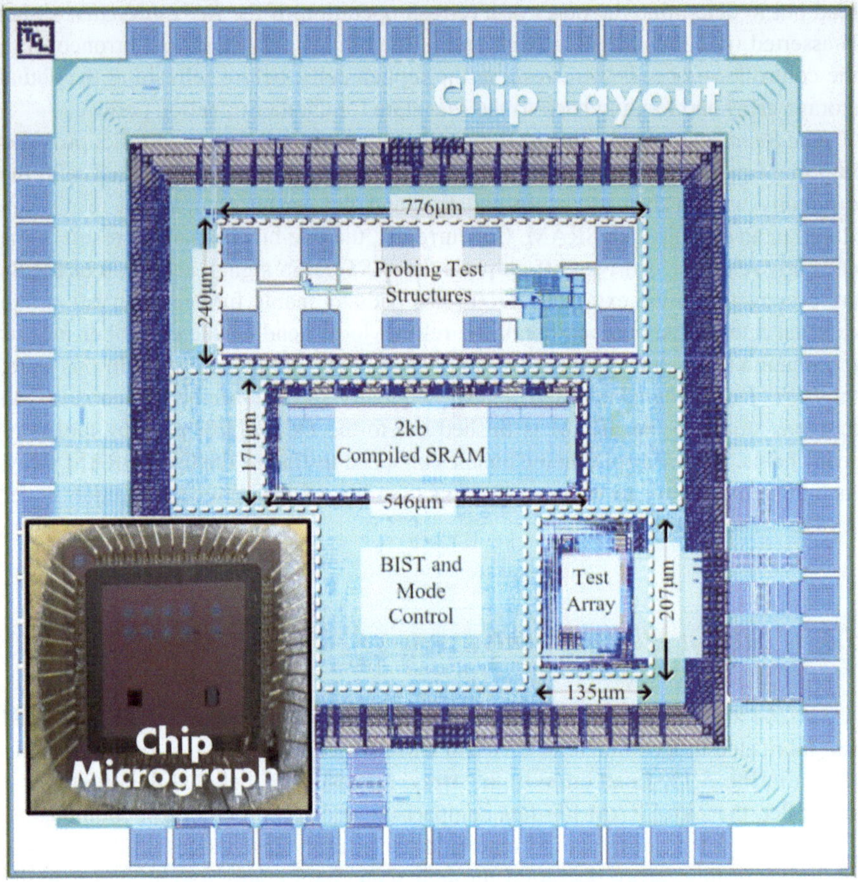

Fig. 5.15 Full layout of the replica GC-eDRAM test chip with major components

with a given frequency of write operations. The figure shows both the mean and minimum retention time of the array. For the shown die, the minimum retention time is lower than the uncalibrated automatic refresh timing. However, the write activity tracking mechanism is shown to work correctly, such that the post-silicon calibration can easily skew the refresh time to a value below the worst-case retention time. This plot emphasizes the efficiency of integrating the replica technique. Traditional worst-case design assumes 100% write activity, resulting in a refresh period of well below 10 ms, even for this typical die. Application of the replica technique adapts this period, refreshing at a more than 5× lower frequency for 10% write activity.

Figure 5.19 shows the dynamic power consumption of the array, as a function of the write and read activity. For a retention period of 20 ms, the active refresh power of the array is 635 fW/bit, which is comparable with previous low-power GC-eDRAM implementations [6].

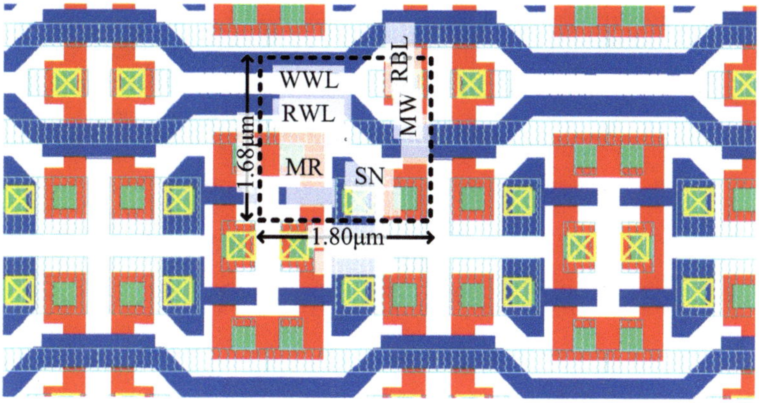

Fig. 5.16 Small section of the GC-eDRAM array layout showing the dimensions of the unit cell

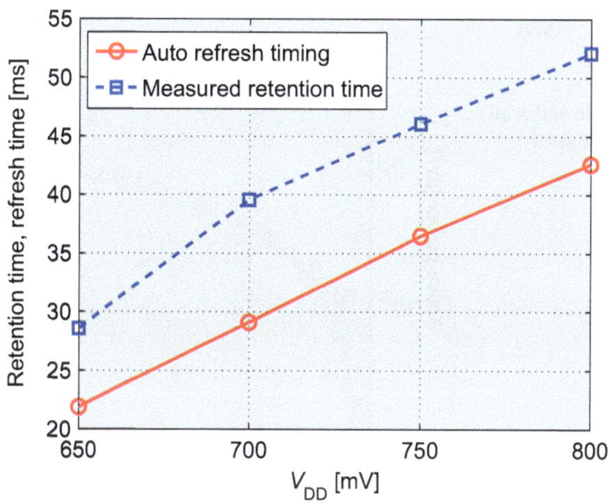

Fig. 5.17 Automatic refresh timing vs. measured retention time for a range of supply voltages

5.5 Conclusions

This chapter presented various bitcell circuit and assist techniques to increase the retention time of NTV GC-eDRAMs, namely a novel TG based 3T GC, reverse body biasing, and replica GCs for retention time tracking. The retention time gain of each technique was reported individually for clarity. While these individual techniques already enable significant retention time boosts, as summarized below, they can also be combined to yield optimized GC-eDRAM designs with data retention power well below the leakage power of 6T SRAM.

Section 5.2 proposed a novel 3T GC-eDRAM macrocell targeted at an energy-efficient NTV VLSI SoC and providing high storage density. The proposed 3T

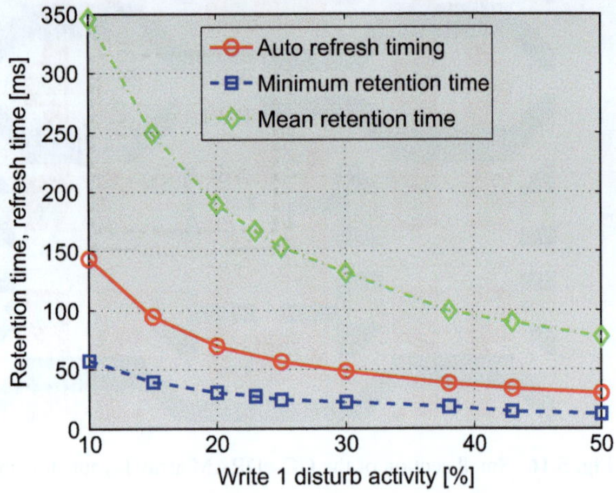

Fig. 5.18 Automatic refresh timing vs. measured retention time for a varying degree of write disturbs

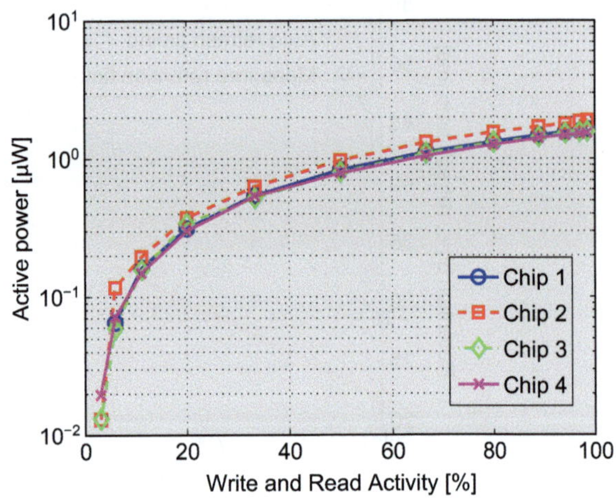

Fig. 5.19 Dynamic power consumption of 2 kb GC-eDRAM array as a function of the write and read activity factor for several measured chips

GC, using a full transmission gate as write port, is operated from a single supply voltage, eliminating the need for boosted voltages commonly found in prior-art GC-eDRAM implementations. The proposed 3T GC exhibits faster write access than conventional GC circuits, while minimizing CI and CF effects, thereby increasing data retention time and reducing refresh power. The cell area is only 57% of a redrawn 6T SRAM in the same technology, making it a suitable alternative to SRAM for energy-efficient memories. A test chip containing a 2 kb memory macro based on the proposed 3T GC was fabricated in a mature 0.18 μm CMOS technology and several chips were tested. Measurement results show full functionality at voltages ranging from 600 mV to 1.8 V with retention power as much as 17× lower than a previously reported 6T SRAM in the same technology node.

Section 5.3 showed the impact of body biasing on the retention time of an all-PMOS 2T GC topology in a mature 0.18 μm CMOS technology. The measured retention time of a 2 kb GC-eDRAM macrocell is improved by 2.3× (from 23 to 53 ms) with a reverse body bias (RBB) of only 100 mV. The cell-to-cell retention time variability is high, ranging from 23 to 569 ms under standard body bias. The absence of a systematic pattern in the measured retention time maps suggests that the high variability is due to local parametric variations, which are particularly high in memory arrays due to the use of minimum-sized devices [4]. Moreover, the process parameters of I/O devices, used to achieve high retention times, may be less carefully controlled than those of core transistors. Nevertheless, RBB is an attractive technique to improve the minimum (as well as the average) retention time.

At the same time, the retention time penalty for FBB (used for fast memory access) is high, exhibiting a 2.9× reduction for 100 mV FBB. However, a possible control scheme could dynamically apply an RBB during retention periods and an FBB during refresh cycles to maximize the array availability. Overall, sweeping the body voltage over a range of 375 mV provides an interesting trade-off between access and retention time, with the retention time range spanning almost 2 orders of magnitude.

In Sect. 5.4, we proposed a replica technique for tracking the PVT variations and operating statistics of a near-V_T GC-eDRAM array for efficient data retention time extension. The technique was implemented on a 2 kb all-PMOS 2T array in a commercial 0.18 μm CMOS process along with an advanced control scheme for extensive testing and measurement. The replica technique was shown to effectively track the retention time of the array across various supply voltages and write activity frequencies, enabling as much as a 5× improvement in retention time, thereby significantly reducing the frequency of power-hungry refresh operations.

References

1. Chun KC, Jain P, Lee JH, Kim C (2011) A 3T gain cell embedded DRAMutilizing preferential boosting for high density and low power on-die caches. IEEE J Solid State Circuits 46(6): 1495–1505
2. Chun KC, Jain P, Kim TH, Kim C (2012a) A 667 MHz logic-compatible embedded DRAM featuring an asymmetric 2T gain cell for high speed on-die caches. IEEE J Solid State Circuits 47(2):547–559. doi:10.1109/JSSC.2011.2168729
3. Chun KC, Zhang W, Jain P, Kim C (2012b) A 2T1C embedded DRAM macro with no boosted supplies featuring a 7T SRAM based repair and a cell storage monitor. IEEE J Solid State Circuits 47(10):2517–2526
4. Ghosh S, Roy K (2010) Parameter variation tolerance and error resiliency: new design paradigm for the nanoscale era. Proc IEEE 98(10):1718–1751
5. Iqbal R, Meinerzhagen P, Burg A (2012) Two-port low-power gain-cell storage array: voltage scaling and retention time. In: Proceedings of the IEEE international symposium on circuits and systems (ISCAS), pp 2469–2472
6. Lee Y, Chen MT, Park J, Sylvester D, Blaauw D (2010) A 5.42 nW/kB retention power logic-compatible embedded DRAM with 2T dual-VT gain cell for low power sensing applications. In: Proceedings of the IEEE Asian solid state circuits conference (A-SSCC), pp 1–4

7. Meinerzhagen P, Teman A, Mordakhay A, Burg A, Fish A (2012) A sub-VT 2T gain-cell memory for biomedical applications. In: Proceedings of the IEEE subthreshold microelectronics conference (SubVT), pp 1–3. doi:10.1109/SubVT.2012.6404318
8. Meinerzhagen P, Teman A, Burg A, Fish A (2013a) Impact of body biasing on the retention time of gain-cell memories. IET J Eng 1:1–4
9. Meinerzhagen P, Teman A, Giterman R, Burg A, Fish A (2013b) Exploration of sub-VT and near-VT 2T gain-cell memories for ultra-low power applications under technology scaling. J Low Power Electron Appl 3(2):54–72. doi:10.3390/jlpea3020054
10. Raychowdhury A, Geuskens B, Kulkarni J, Tschanz J, Bowman K, Karnik T, Lu SL, De V, Khellah M (2010) PVT-and-aging adaptive wordline boosting for 8T SRAM power reduction. In: Proceedings of the IEEE international solid-state circuits conference (ISSCC), pp 352–353
11. Somasekhar D, Ye Y, Aseron P, Lu SL, Khellah M, Howard J, Ruhl G, Karnik T, Borkar S, De V, Keshavarzi A (2009) 2 GHz 2 Mb 2T gain cell memory macro with 128 GBytes/sec bandwidth in a 65 nm logic process technology. IEEE J Solid State Circuits 44(1):174–185
12. Sreerama Reddy G et al (2009) Design and implementation of 8k-bits low power SRAM in 180nm technology. In: Proceedings of the IMCECS, vol 2
13. Teman A, Meinerzhagen P, Burg A, Fish A (2012a) Review and classification of gain cell eDRAM implementations. In: Proceedings of the IEEE convention of electrical and electronics engineers in Israel (IEEEI), pp 1–5
14. Teman A, Yadid-Pecht O, Fish A (2012b) Leakage reduction in advanced image sensors using an improved AB2C scheme. IEEE Sensors J 12(4):773–784
15. Teman A, Meinerzhagen P, Giterman R, Fish A, Burg A (2014) Replica technique for adaptive refresh timing of gain-cell embedded DRAM. IEEE Trans Circuits Syst II 61(4):259–263
16. Zhang W et al (2013) A write-back-free 2T1D eDRAM with local voltage sensing and a dual-row-access LP mode. IEEE Trans Circuits Syst I 60(8):2030–2038

Chapter 6
Aggressive Technology and Voltage Scaling (Down to the Subthreshold Domain)

6.1 Introduction

Many ultra-low power (ULP) systems, such as biomedical sensor nodes and implants, are required to operate with extremely low power budgets in order to run on a single cubic-millimeter battery charge for days or even for years. Aggressive supply voltage (V_{DD}) scaling, leading to near-threshold (near-V_{T}) or even to sub-V_{T} circuit operation, is widely used in this context to lower both active energy dissipation and leakage power consumption; albeit, at the price of severely degraded on/off current ratios ($I_{\mathrm{on}}/I_{\mathrm{off}}$) and increased sensitivity to process variations [11]. The majority of these biomedical systems require a considerable amount of embedded memory for data and instruction storage, often amounting to a dominant share of the overall silicon area and power. Typical storage capacity requirements range from several kb for low-complexity systems [6] to several tens of kb for more sophisticated systems [5].

Over the last decade, robust, low-leakage, low-power sub-V_{T} memories have been heavily researched [1, 9, 13, 14, 16]. In order to guarantee reliable operation in the sub-V_{T} domain, many new SRAM bitcells consisting of 8 [3, 12], 9 [13, 17], 10 [1], and up to 14 [6] transistors have been proposed. These bitcells utilize the additional devices to solve the problems of write contention and read disturb, and, in addition, some of the designs reduce leakage by using transistor stacks. However, all these state-of-the-art sub-V_{T} memories are based on static bitcells, not taking into consideration the applicability of dynamic memories for this type of operation.

This chapter considers the design and operation of gain-cell embedded DRAMs (GC-eDRAMs) at aggressively scaled supply voltages and under aggressive technology scaling. We show that the current ratios are still high enough in the sub-V_{T} regime to achieve short access and refresh cycles and high memory availability, at least down to 0.18 µm CMOS nodes. A GC-eDRAM array, which targets aggressively scaled supply voltages, down into the subthreshold domain is proposed. Minimum supply voltage (V_{DD}) design of gain-cell arrays is evaluated in light of

P. Meinerzhagen et al., *Gain-Cell Embedded DRAMs for Low-Power VLSI Systems-on-Chip*, DOI 10.1007/978-3-319-60402-2_6

technology scaling, considering both a mature 0.18 µm CMOS node as well as a scaled 40 nm node. We first analyze the trade-offs that characterize the bitcell design in both nodes, presenting a best-practice design methodology for both mature and scaled technologies. Following this analysis, we propose full gain-cell arrays for both considered nodes, operated at a minimum V_{DD}. We find that an 0.18 µm CMOS gain-cell array can be robustly operated at a sub-V_T supply voltage of 400 mV with read/write availability over 99% of the time, despite refresh cycles. This is demonstrated on a 2 kb array, operated at 1 MHz, and exhibiting full functionality under parametric variations. As opposed to the mature node, we find that the scaled 40 nm node cannot be operated in sub-V_T, and instead requires a near-threshold 600 mV supply to achieve at least 97% read/write availability. This limitation is primarily due to higher leakage currents that limit the retention time of the bitcell. Monte Carlo (MC) simulations show that a 600 mV 2 kb 40 nm gain-cell array is fully functional at frequencies higher than 50 MHz.

To summarize, GC-eDRAMs implemented in mature CMOS nodes can successfully be operated at aggressively scaled sub-V_T voltages, whereas voltage scaling for GC-eDRAM implementations in aggressively scaled CMOS nodes is best limited to the near-V_T domain.

6.2 Retention Time Model Validation for 28 nm CMOS

In Chap. 3, a model for the effective data retention time (EDRT) of a 2-transistor (2T) GC-eDRAM bitcell was developed and silicon proven for a mature 0.18 µm CMOS technology. The consideration of such a mature node was targeted at the implementation of ultra-low power biomedical applications; however, the concepts shown in the proposed model are generic and applicable under technology scaling. To demonstrate this, the 2T GC-eDRAM circuit was implemented in a state-of-the-art 28 nm CMOS process technology, using a high-V_T write transistor (MW) and a standard-V_T read transistor (MR) from the technology. In addition to the simulations performed to extract the model parameters of (3.7) (see Chap. 3), extensive leakage simulations were carried out, showing that for this technology, the sub-V_T leakage through MW is still the dominant factor that limits the retention time of a 2T GC-eDRAM bitcell, constituting over 97% of the leakage into the storage node (SN).

To evaluate the matching of the proposed model to the statistical distribution of EDRT in the chosen 28 nm technology, 10,000 MC bitcell simulations were carried out. The resulting distribution is shown in Fig. 6.1. This figure superimposes the close match of the proposed model to the distribution, showing that the model still provides a good estimation of the expected distribution, and that the exponential dependence of retention time on the threshold voltage of the write transistor (V_T^{MW}) still holds. Furthermore, worst-case design (WCD) high-yield analysis was performed on the extreme values of the model, and a matching within 10% was found, which is entirely acceptable for the purposes of the proposed model.

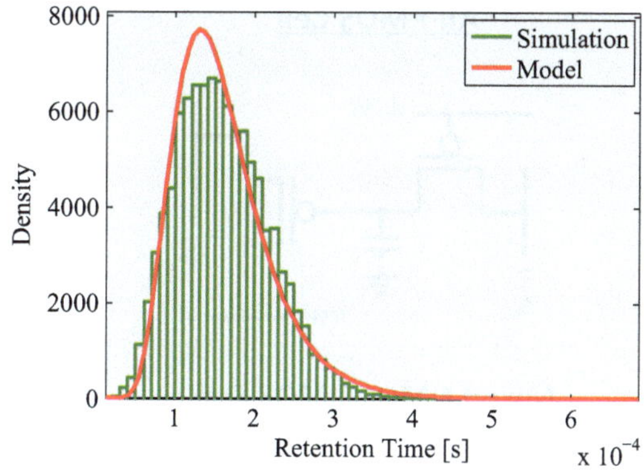

Fig. 6.1 Statistical EDRT distribution: model and MC simulation in 28 nm CMOS

6.3 2T Gain Cells Optimized for Subthreshold Operation

Previously reported gain-cell topologies include either two or three transistors and an optional MOSCAP [15]. While the basic 2T bitcell has the smallest area cost, it limits the number of cells which can connect to the same read bitline (RBL) due to leakage currents from unselected cells masking the sense current [8]. However, as many ULP systems require only small memory arrays with relatively few cells per RBL, in this section, we consider the implementation of a 2T bitcell as a viable low-voltage option and propose a best-practice 2T bitcell design for the considered technology nodes (0.18 µm and 40 nm).

6.3.1 2T Gain-Cell Implementation Alternatives

Figure 6.2 shows the four basic options for implementing a 2T gain-cell, allowing both MW and MR to be implemented with either an NMOS or a PMOS device. These standard topologies require the following control schemes to achieve robust write and read operations. A boosted write wordline (WWL) voltage is required during write access due to V_T drop across MW. This boosted voltage needs to be above V_{DD} for the NMOS option (V_{BOOST}) and below V_{SS} for the PMOS option (V_{NWL}). For a read operation with a PMOS MR, the parasitic RBL capacitance is pre-discharged, and the read wordline (RWL) is subsequently raised. If the storage node (SN) of the selected bitcell holds a "0," MR is turned-on, and therefore, charges RBL past a detectable sensing threshold. If SN holds a "1," MR is cut off, and therefore, RBL remains discharged below the sensing threshold. Using an NMOS transistor to implement MR provides the exact opposite operation, i.e., RBL is pre-charged and RWL is lowered to initiate a read.

In the mature 0.18 µm CMOS technology, both MW and MR can be implemented with either standard-V_T (SVT) core or high-V_T (HVT) I/O devices. In more advanced technology nodes, typically starting with the 130 nm or 90 nm node

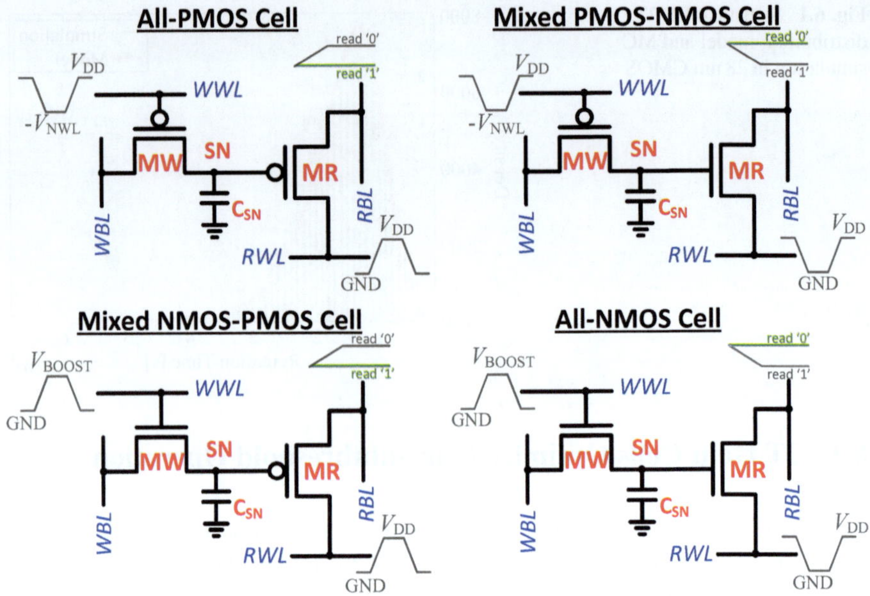

Fig. 6.2 2T gain-cell implementation options including the schematic waveforms

for most semiconductor foundries, several V_T options become available for core devices, most commonly low-V_T (LVT), SVT, and HVT devices. One of the primary requirements of gain-cell implementation is to achieve sufficiently high retention times to provide high array availability and reduce the energy spent on refresh operations.

As previously discussed, the retention time is the time it takes for the level stored on SN to deteriorate through leakage currents. In mature, above-100 nm CMOS nodes, sub-V_T conduction is the dominant leakage mechanism, compromising data retention in any 2T gain-cell through the channel of MW, as shown in Fig. 6.3a. Therefore, the primary selection criterion for the device type of MW is to minimize sub-V_T conduction. Note that sub-V_T conduction of MW weakens both a logic "1" and a logic "0" level, whenever the write bitline (WBL) voltage is opposite to the SN voltage.

In more advanced, sub-100 nm CMOS nodes, there are other significant leakage mechanisms that can compromise data integrity.[1] Only leakage components that bring charge onto the SN or remove charge from SN need to be considered in terms of retention time, while other leakage components are merely undesirable in terms of static power consumption. Figure 6.3b schematically shows the main

[1]Note that in the sub-V_T region, these mechanisms are still negligible, as compared to sub-V_T conduction. However, as shown in Sect. 6.3.3.2, at near-V_T supplies, some of the mechanisms still must be considered.

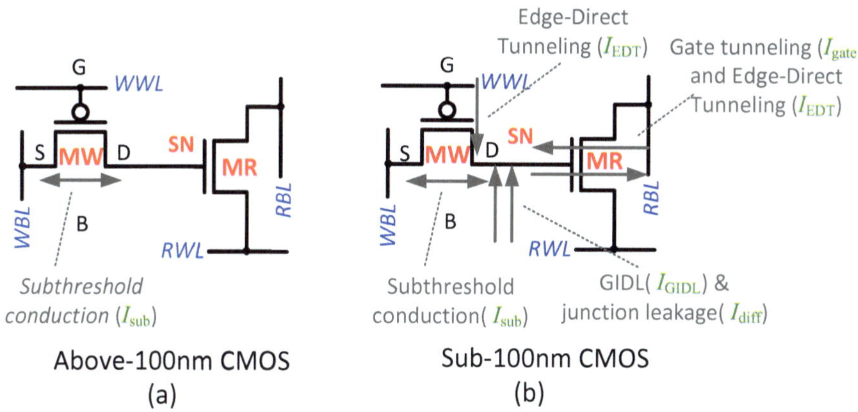

Fig. 6.3 Leakage components which are considered for the choice of the best-practice write and read transistor implementations, for (**a**) mature CMOS nodes, and (**b**) scaled CMOS nodes

leakage components that can compromise the stored level in sub-100 nm nodes. These include reverse-biased pn-junction leakage (I_{diff}), gate-induced drain leakage (I_{GIDL}), gate tunneling leakage (I_{gate}), edge-direct tunneling current (I_{EDT}), and sub-V_T conduction (I_{sub}). When employing a PMOS MW, the bulk-to-drain leakages (I_{diff} and I_{GIDL}) weaken a logic "0" and strengthen a logic "1," but have the opposite impact (i.e., strengthen a logic "0" and weaken a logic "1") when MW is implemented with an NMOS device. During standby, MW is always off and has no channel; therefore, forward gate tunneling (I_{gate}) from the gate into the channel region and into the two diffusion areas that would occur in a turned-on MOS device is of no concern here. Only the edge-direct tunneling current, from the diffusion connected to the SN in the absence of a strongly inverted channel, compromises data integrity. When implementing MW with an NMOS device, edge-direct tunneling discharges a logic "1," while it charges a logic "0" for a PMOS MW.

The only terminal of MR that is connected to SN is the gate terminal, and therefore, the only leakage through MR that affects the stored data level is gate tunneling. During standby, there is no channel formation in MR, no matter what the stored data level is. For example, if using an NMOS MR, both RWL and RBL are charged to V_{DD} during standby, such that even a logic "1" level results in zero gate overdrive. In this case, both diffusion areas of MR are at the same potential as the SN, eliminating tunneling currents between the diffusions and the gate ($I_{EDT} = 0$). However, tunneling may occur from the gate directly into the grounded bulk (I_{gate}), weakening a logic "1." If the same cell stores a logic "0," tunneling between the gate and bulk is avoided ($I_{gate} = 0$), while reverse tunneling from the diffusions (I_{EDT}) into the gate can charge the logic "0" level. The exact opposite biasing conditions and corresponding tunneling mechanisms are found when MR is implemented with a PMOS.

6.3.2 Best-Practice Write Transistor Implementation

6.3.2.1 Mature 0.18 μm CMOS Node

For ULP sub-V_T applications, long retention times that minimize the number of power-consuming refresh cycles are of much higher importance than fast write access. Therefore, low sub-V_T leakage becomes the primary factor in the choice of a best practice write transistor in the 0.18 μm node. The off current (I_{off}) of NMOS and PMOS, core and I/O devices offered in this process are shown in Fig. 6.4a. Clearly, the I/O PMOS device has the lowest I_{off} ($V_{GS} = 0$ V, $V_{DS} = -V_{DD}$) among all device options and across all standard process corners, leading to the longest retention time. At a 400 mV sub-V_T V_{DD}, the on-current, I_{on} ($V_{GS} = -V_{DD}$, $V_{DS} = -V_{DD}$), of this preferred I/O PMOS device is still four orders-of-magnitude larger than I_{off}, as shown in Fig. 6.4b, which results in sufficiently fast write and refresh operations compared to the achievable retention time. This holds for temperatures up to 37 °C, which is considered a maximum, worst-case temperature for ULP systems targeted at biomedical applications that are typically attached to the human body and hardly suffer from self-heating due to low computational complexity. Nevertheless, for temperatures as high as 125 °C, a sufficiently high I_{on}/I_{off} ratio of four orders-of-magnitude is still achieved at a slightly higher V_{DD} of 500 mV.

Figure 6.5a shows the worst-case data deterioration as a function of time after writing into a 2T gain-cell with a PMOS I/O write transistor under global and local variations. The blue (bottom) curves show the deterioration of a logic "0" level with WBL tied to V_{DD}, and the red (top) curves show the deterioration of a logic "1" level with WBL tied to ground. The plot was extracted for a sub-V_T 400 mV V_{DD} assuming a storage node capacitance of 2.5 fF. A worst-case retention time of 40 ms can be estimated from this figure, by finding the minimum time at which the "0" and "1" levels intersect. It is clear that logic "0" levels decay much faster than logic "1" levels, corresponding with previous reports for the above-V_T domain [4, 7]. In fact, the decay of a "1" level is self-limited due to the steady increase of the reverse gate overdrive ($V_{GS,MW} = V_{DD} - V_{SN}$) and the increasing body effect ($V_{BS,MW} = V_{DD} - V_{SN}$) of MW with progressing decay. Both of these effects suppress the leakage of the device, resulting in slower level decay. Furthermore, the charge injection (CI) and clock feedthrough (CF) that occur at the end of a write access (when MW is turned off) cause the SN voltage level to rise, thereby strengthening a "1" and weakening a "0" level [10, 15]. Therefore, careful consideration must be given to the initial state of the "0" level following a write access, as will be discussed in Sect. 6.3.4.

6.3.2.2 Scaled 40 nm CMOS Node

To achieve minimum V_{DD} operation at the scaled 40 nm node, sub-V_T conduction must again be kept as small as possible when choosing the best device option for MW, as it affects both a "1" and a "0" level. In contrast, the diffusion leakage, the

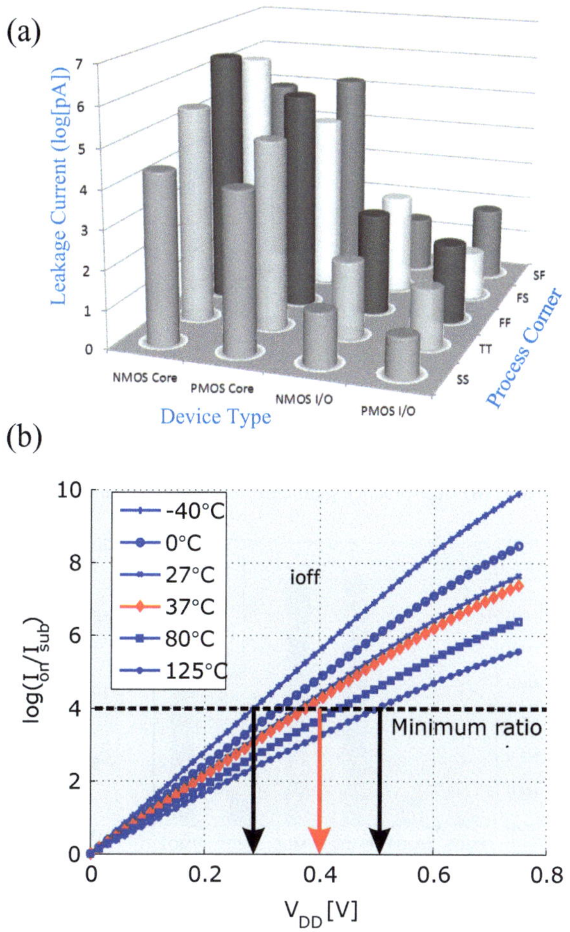

Fig. 6.4 (**a**) Sub-V_T conduction of different transistor types in an 0.18 µm node, and (**b**) I/O PMOS I_{on}/I_{off} current ratio as a function of V_{DD} for the typical-typical (TT) process corner at different temperatures

GIDL current, and the edge-direct tunneling current weaken one logic level, while strengthening the other. However, all three leakage components work against the logic level which has already been weakened through CI and CF at the end of a write pulse. For example, a positive SN voltage step occurs when closing a PMOS MW, thereby weakening the logic "0" level. Unfortunately, I_{GIDL}, I_{diff}, and I_{EDT} also pull up SN, further deteriorating the stored "0." Therefore, in order to protect the already weaker level, the optimum device selection aims at minimizing all of these leakage components. Figure 6.6a shows the leakage components of minimum sized devices provided in the considered 40 nm process[2] at a near-V_T supply voltage of 600 mV. This figure clearly shows that despite the increasing significance of

[2]The LVT devices were left out of the figure for display purposes, as their leakage is significantly higher than the leakage of other devices.

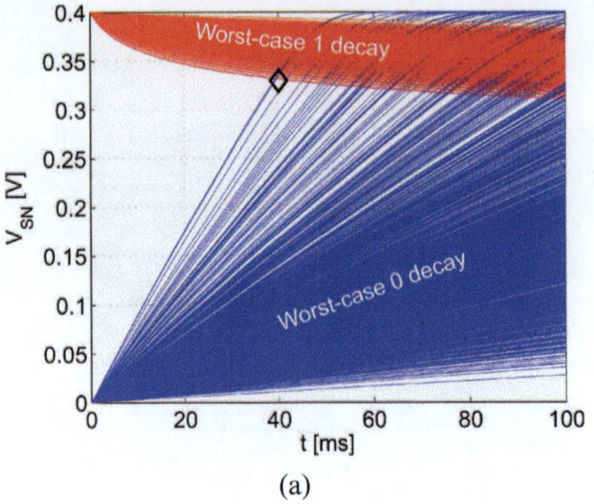

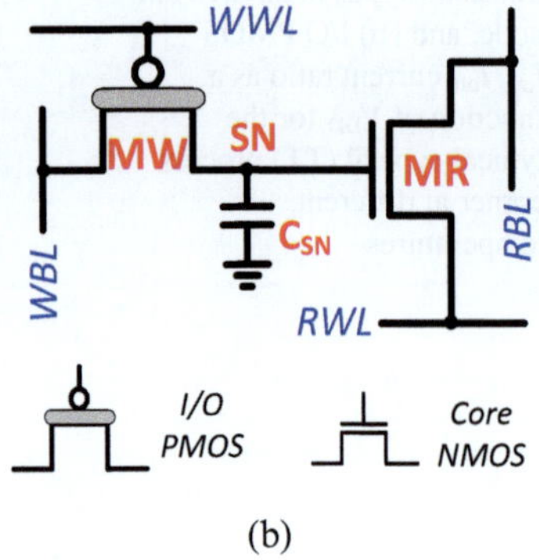

Fig. 6.5 (**a**) Worst-case retention time estimation of 0.18 µm sub-V_T gain-cell with V_{DD} = 400 mV. (**b**) Best-practice gain-cell for sub-V_T operation in 0.18 µm CMOS

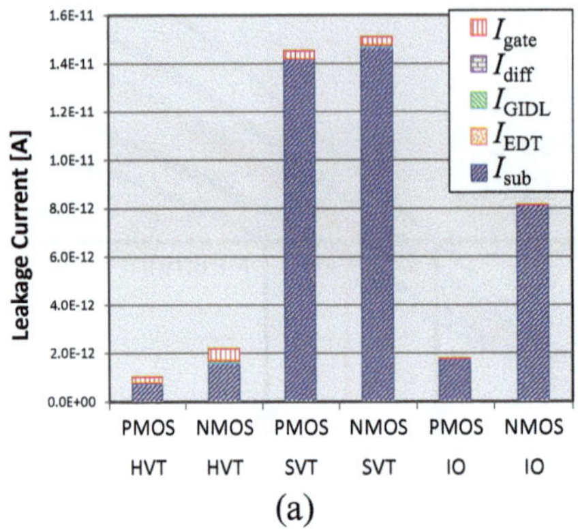

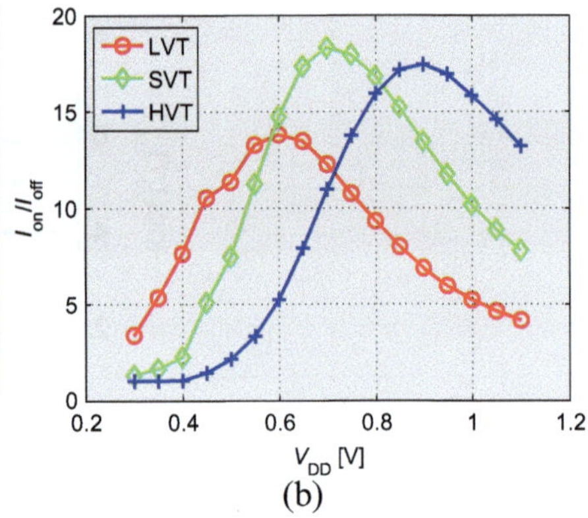

Fig. 6.6 (**a**) Leakage components of various devices in the considered 40 nm node at a near-V_T supply voltage of 600 mV. (**b**) Worst-case I_{on}(weak "1")/I_{off}(weak "0") of MR, implemented with LVT, SVT, and HVT devices. Both plots were simulated under typical conditions

other leakage currents with technology scaling, sub-V_T conduction (I_{sub}) is still dominant at this node.[3] However, the advantage of using an I/O device is lost, and a more compact HVT PMOS device provides the lowest total leakage. This trend is confirmed when evaluating the leakage components of intermediate process nodes and further nodes down to 28 nm, as well. In fact, the leakage benefits of using an I/O device deteriorate to the point where the area versus leakage trade-off favors the use of an HVT device at around the 65 nm node.

[3]Some of the leakage components are not modeled for the I/O devices; however, this does not impact our analysis, as the PMOS HVT already provides the lowest total leakage.

6.3.3 Best-Practice Read Transistor Implementation

6.3.3.1 Mature 0.18 μm CMOS Node

At the onset of a read operation, RWL is asserted (driven high for an NMOS MR and low for a PMOS MR), and capacitive coupling from RWL to SN causes a voltage step on SN [10]. Our analysis from the previous section showed that MW should be implemented with a PMOS device, resulting in a strong logic "1" and a weaker logic "0." Therefore, it is clearly preferable to implement MR with an NMOS transistor that employs a negative RWL transition for read assertion. The resulting decrease in voltage on SN counteracts the previous effects of CI and CF, thus improving the "0" state during a read operation. Note that this change is temporary, as during de-assertion of RWL, the opposite coupling occurs. As a side effect, the negative SN voltage step also lowers the "1" level, and therefore slightly slows down the read operation; however, this level is already initially boosted due to de-assertion of the WWL. An additional, and perhaps more significant reason to choose an NMOS device for readout is that NMOS devices are approximately an order-of-magnitude stronger than their PMOS counterparts at sub-V_T voltages. Therefore, implementing MR with an NMOS device provides a fast read access, which not only results in better performance, but is essential for ensuring high array availability. As mentioned, the considered 0.18 μm process includes core and I/O devices, with the core devices providing approximately three-orders-of-magnitude higher on-current at sub-V_T voltages. Therefore, the choice of implementing MR with a core NMOS is straightforward.

To summarize, the most appropriate 2T gain-cell for sub-V_T operation in an above-100 nm CMOS node comprises an I/O PMOS write transistor and a core NMOS read transistor. This configuration is illustrated in Fig. 6.5b. The resulting hybrid NMOS/PMOS gain-cell shares the n-well on three sides between neighboring cells to keep the area cost low [8]. This is further discussed in Sect. 6.4.

6.3.3.2 Scaled 40 nm CMOS Node

When considering the best device type for scaled nodes, the large number of options presents some interesting trade-offs for the implementation of MR. Scaled nodes are characterized by increasing gate leakage currents (I_{gate} and I_{EDT}), which could potentially present an advantage for a thick oxide I/O device due to its reduced gate tunneling. However, at low voltages, the tunneling currents are small in comparison with the sub-V_T conduction through MW. This is shown in Fig. 6.6a. In addition, I_{gate} and I_{EDT} actually appear in opposite directions, as the stored "0" level rises, further reducing their impact.

On the other hand, the two primary considerations for the above-100 nm nodes are even more relevant at scaled nodes. The achievable retention time in the 40 nm process is approximately three orders-of-magnitude lower than that of the 0.18 μm node. Therefore, the negative step caused by RWL coupling to SN is even more important, and with the required high refresh rates, fast reads are essential to provide sufficient array availability. To further enhance the step during RWL assertion, layout techniques can be implemented to increase the capacitive coupling between RWL and SN.

However, when considering read access times, additional trade-offs arise. For maximum read performance, MR could be implemented with an LVT NMOS device, which provides an 8× increase in on-current at 400 mV compared to an SVT NMOS. As the supply voltage is increased, this benefit reduces to 3× at 600 mV. The superior on-currents of LVT devices, as compared to SVT or HVT options, come at the expense of much higher off-currents, as well as increased process variations.

When choosing the read device, this trade-off must be taken into consideration, as it is mandatory to correctly differentiate between the discharged level of RBL due to a stored "1" and the depleted level due to a weak stored "0." Furthermore, the unselected cells on the same column of a selected cell storing a "1" will start to counteract the discharge of RBL during a read, as $V_{GS,MR}^{unselected} = V_{DD} - V_{RBL}$. This phenomenon is referred to as "RBL Saturation." In effect, this limits the speed and minimum discharge level of RBL, according to the drive strength of the unselected MR devices. When considering sub-V_T operation in the 40 nm node, the relatively low sub-V_T conduction of the SVT, HVT, and I/O devices, render the LVT the only feasible option for MR to achieve a reasonable RBL discharge time. However, as V_{DD} is increased into the near-V_T region, an SVT device provides sufficient on-current, while the higher V_T and lower leakage provide better reliability under process variations, as well as improved array availability.

Figure 6.6b shows the worst-case current ratio I_{on}/I_{off} of the NMOS read transistor MR, implemented with different device types as a function of V_{DD}. I_{on} is given for a weak "1" level, estimated as the steady-state high voltage of SN when tying WBL to V_{DD} ($V_{SN} = 0.85V_{DD}$). I_{off} is given for a weak "0" level, estimated at $V_{SN} = 0.4V_{DD}$, which would provide a sufficient margin to differentiate between the two levels.[4] For supply voltages below 600 mV, the highest current ratio is provided by the LVT device, leading to the best achievable array availability. Therefore, implementing MR with an LVT NMOS is the preferred choice. Likewise, the SVT device is preferred for V_{DD} between 600 and 800 mV, while the HVT device is the best option for even higher values of V_{DD}.

6.3.4 Storage Node Capacitance and WWL Underdrive Voltage

6.3.4.1 Mature 0.18 μm CMOS Node

To close the design of the 2T bitcell, two important design parameters must be taken into consideration. First, the storage node capacitance (C_{SN}), primarily made up of the diffusion capacitance of MW and the gate capacitance of MR, is typically around

[4]This is verified for the chosen implementation at the minimum feasible bias in Sect. 6.5.

1 fF for minimum device sizes. However, by applying layout techniques, such as metal stacking, this value can be extended by over 5×, providing a configurable design parameter. Second, to address the V_T drop across MW that primarily affects the write "0" operation (but also the write "1" operation in the sub-V_T regime), an underdrive voltage (V_{NWL}) needs to be applied to WWL. The magnitude of V_{NWL} affects both the write access time and the SN voltage.

Figure 6.7a shows the storage node voltage (V_{SN}) after a write "0" access as a function of C_{SN} and V_{NWL}, before and after closing MW. The impact of CI and CF is emphasized in Fig. 6.7b by showing the voltage step ΔV that occurs while closing MW. It is clear that any V_{NWL} above −650 mV already results in a degraded logic "0" transfer prior to turning off MW. ΔV can be reduced by increasing C_{SN} and by decreasing the magnitude of V_{NWL}. Therefore, on the one hand, V_{NWL} must be low enough to ensure a proper transfer of logic "0," but, on the other hand, it should be as high as possible to minimize ΔV. The optimum value for V_{NWL} leading to the

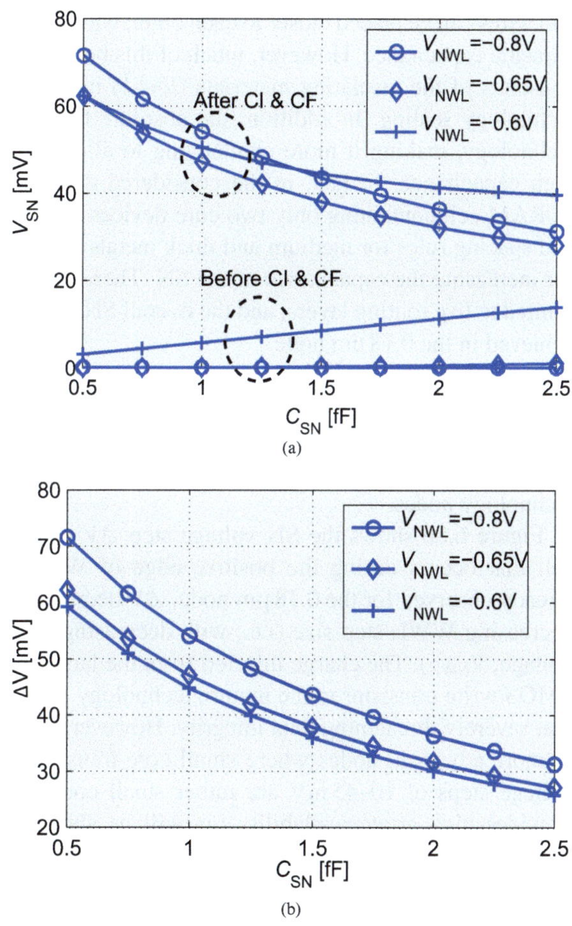

Fig. 6.7 Following a write "0" operation: (**a**) V_{SN} before and after closing MW, as a function of C_{SN} and V_{NWL}. (**b**) ΔV due to charge injection from MW and due to capacitive coupling from WWL to SN

strongest "0" state after a completed write operation is found to be −650 mV, as shown in Fig. 6.7a. The optimum value for C_{SN} is clearly the maximum displayed value of 2.5 fF.

6.3.4.2 Scaled 40 nm CMOS Node

It is clear that the storage node capacitance should always be as large as possible, regardless of the technology node. This not only results in an improved initial "0" level, as discussed above, but also provides more stored charge and thus extends the retention time. A general characteristic of scaled CMOS nodes is the increased number of routing layers, which can be used to build up the storage node capacitor through metal stacking. Here, we assume that all available metal layers can be used at no additional cost, as the memory is going to be embedded in a system-on-chip which uses a large metal stack, in any case. Moreover, with technology scaling, the aspect ratio of metal wires changes to narrower but higher, and wires can be placed closer to each other, which is beneficial in terms of side-wall parasitic capacitance. However, much of this benefit is offset by the lower dielectric constants of the insulating materials (*low-k*) integrated into digital processes with technology scaling. In addition, the absolute footprint of the bitcell shrinks with technology, making it more challenging to allocate many inter-digit fingers for a high capacitance. In fact, in the considered 40 nm node, the footprint of a GC-eDRAM cell containing only two core devices is so small that the minimum width and spacing rules for medium and thick metals are too large to exploit these metals for increasing the capacitance of the SN. Therefore, our layout of the 40 nm cell is limited to five routing layers, and the overall SN capacitance is much lower than that achieved in the 0.18 μm node.

Note that all of our layout analysis assumes adherence to general purpose design rules, which in some cases could be waived in order to achieve improved characteristics. Figure 6.8a summarizes the achievable storage node capacitance according to the number of thin metal layers provided by the two considered technology nodes.

Figure 6.8b shows the SN voltage step ΔV of the 40 nm CMOS GC-eDRAM cell that occurs during the positive edge of WWL for a logic "0" transfer. As already observed for the 0.18 μm node, ΔV decreases with increasing C_{SN} and with decreasing WWL step size (i.e., with decreasing absolute value of the underdrive voltage, V_{NWL}). The charge injected from the large channel area of the selected I/O PMOS write transistor in the mature technology node results in a large voltage step that severely threatening data integrity. However, this problem is slightly alleviated in more advanced nodes where small core transistors are preferred. The resulting voltage steps of 10–45 mV are rather small compared to the minimum V_{DD} that provides high array availability (as will be shown in Sect. 6.5). Moreover, it is worth mentioning that strong "0" levels are transferred to SN even with the least

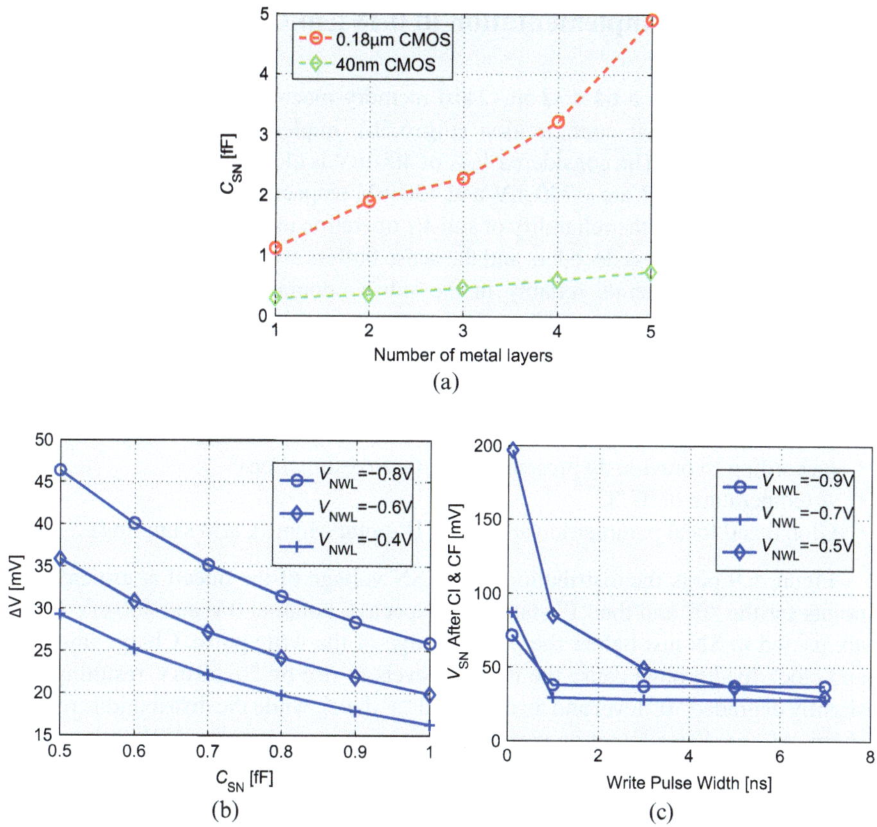

Fig. 6.8 (**a**) Storage node capacitance versus number of employed metal layers. (**b**) ΔV due to CI and CF, as a function of C_{SN} and V_{NWL}, for $V_{DD} = 700\,\text{mV}$. (**c**) V_{SN} after CI and CF versus write pulse width

aggressive underdrive voltage of −0.4V (however, at the expense of write access time). Therefore, the ΔV values in Fig. 6.8b also correspond to the final SN voltage immediately following the write access.

The final choice of V_{NWL} for the 40 nm node needs to account for the write access time, which must remain short to guarantee high array availability in a node with high leakage and short retention times (see Sect. 6.5). Therefore, Fig. 6.8c shows the final V_{SN} after CI and CF, as a function of the write pulse width. Over a large range of pulse widths as short as several nanoseconds, an underdrive voltage of −700 mV results in the strongest "0" levels, and is therefore preferred. Less underdrive (e.g., −500 mV) would result in weak "0" levels for pulse widths that are shorter than 3 ns.

6.4 Macrocell Implementation in 0.18 μm CMOS

This section presents a 64 × 32 bit (2 kb) memory macro based on the previously described 2T gain-cell configuration (Fig. 6.5b), implemented in a bulk CMOS 0.18 μm technology. The considered V_{DD} of 400 mV is clearly in the sub-V_T regime, as V_T of MW and MR are −720 mV and 430 mV, respectively. Special emphasis is put on the analysis of the reliability of sub-V_T operation under parametric variations.

Note that the address decoders and the sense buffers are built from combinational CMOS gates and operate reliably in the sub-V_T domain [2], and therefore, the analysis focuses on the write-ability, data retention, and read-ability of the gain-cell. All simulations assume the following test setup:

- A 1 μs write and read access time (1 MHz operation)
- A 3-metal SN capacitance of 2.5 fF, providing a retention time of 40 ms (according to previously presented worst-case estimation)
- A temperature of 37 °C
- Global and local parametric variations (1k-point Monte Carlo sampling).

Figure 6.9 plots the distribution of the SN voltage of the bitcell at critical time points for the "0" and the "1" states. As expected, nominal 0 V and 400 mV levels are passed to SN just before the positive edge of the write pulse. Charge injection and clock feedthrough cause the internal levels to rise by 20–50 mV, resulting in a slightly degraded "0" level and an enhanced "1" level, while the distributions remain sharp. After a 40 ms retention period with a worst-case opposite WBL voltage, the distributions are spread out, but the "1" levels are still strong. The extreme cases of the "0" levels have severely depleted, approaching 200 mV. However, the "0"

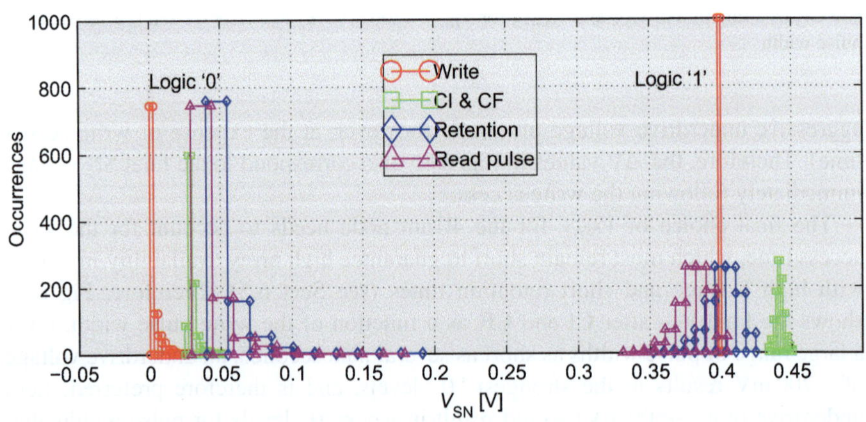

Fig. 6.9 Distribution of the SN voltage of a logic "0" and a logic "1" at critical time points: (1) [*circles*] directly after a 1 μs write access (before turning off MW); (2) [*squares*] after turning off MW; (3) [*diamonds*] after a 40 ms retention period under worst-case WBL conditions; and (4) [*triangles*] during a read operation

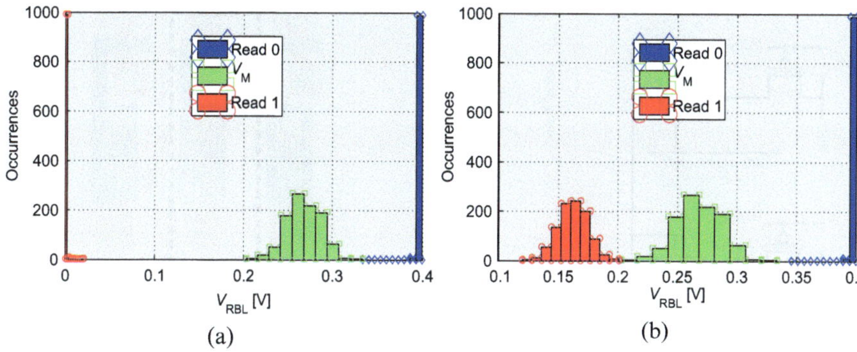

Fig. 6.10 Distribution of RBL voltage (V_{RBL}) after read "1" [*circles*] and read "0" [*diamonds*] operations and distribution of the trip-point V_M of the read buffer [*squares*], for (**a**) favorable and (**b**) unfavorable read "1" conditions

and "1" levels are still well separated, and moreover, the "0" levels are improved following the falling RWL transition, resulting in a 10–20 mV decrease.

To verify the read-ability of the bitcell, Fig. 6.10 shows the distribution of the RBL voltage (V_{RBL}) following read "0" and read "1" operations after the 40 ms retention period. In addition, the figure plots the distribution of the switching threshold (V_M) of the sense buffer. While read "0" is robust in all cases (RBL stays precharged), read "1" is most robust if all unselected cells on the same RBL as the selected cell store "0" (see Fig. 6.10). Read "1" becomes more critical if all unselected cells store "1" (see Fig. 6.10), as this inhibits the discharge of RBL through the selected cell. This worst-case scenario for a read "1" operation is illustrated in Fig. 6.11. In order to make the read operation more robust, V_M is shifted to a value higher than $V_{DD}/2$ by appropriate transistor sizing in the sense inverter. Ultimately, the V_{RBL} distributions for read "0" and read "1" are clearly separated, and the distribution of V_M is shown to fit comfortably between them, as shown in Fig. 6.10.

Figure 6.11b show the layout of the 0.18 μm 2T gain-cell, comprising a PMOS I/O MW and an NMOS core MR. The figure presents a zoomed-in view of one bitcell (surrounded by a dashed line) as part of an array. The chosen technology incurs rather harsh design rules for the implementation of I/O devices; however, by sharing the n-well on three sides and stacking the bitlines, a reasonable area of 4.35 μm^2 per bitcell is achieved. For comparison, a single-ported 6T SRAM bitcell, designed for above-V_T operation in the same node, has a comparable area cost of 4.1 μm^2 (cell violates standard DRC rules). Furthermore, SRAM bitcells optimized for robust operation at low voltages are clearly larger (e.g., the 14T SRAM bitcell in [6] has an area cost of 40 μm^2). The layout of Fig. 6.11b also enables metal stacking above the storage node to provide an increased SN capacitance of up to 5 fF (see Fig. 6.8a).

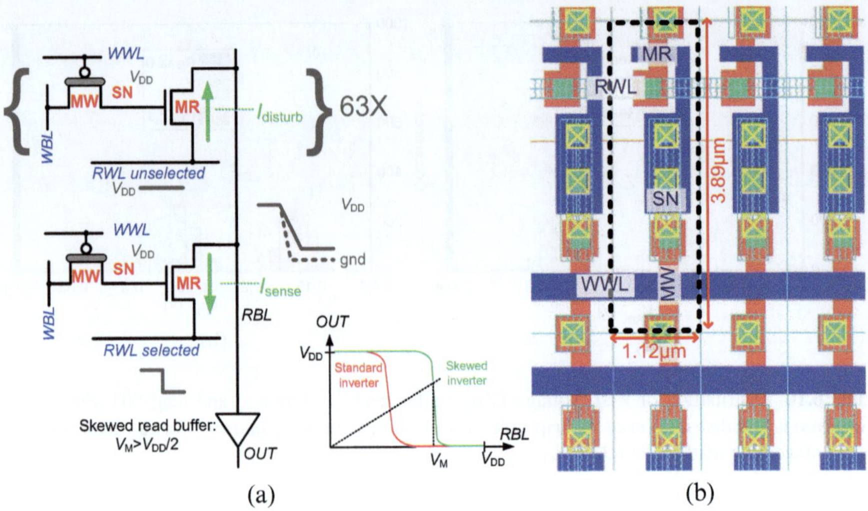

Fig. 6.11 180 nm gain-cell array: (**a**) Worst-case for read "1" operation: all cells in the same column store data "1." To make the "1" operation more robust, the sense inverter is skewed, with a trip-point $V_M > V_{DD}/2$. (**b**) Zoomed-in layout

At an operating frequency of 1 MHz, a full refresh cycle of 64 rows takes approximately 128 µs. With a worst-case 40 ms retention time, the resulting availability for write and read is 99.7%, which is sufficient for most applications. As summarized in Table 6.1, the average leakage power of the 2 kb array at room temperature (27 °C) is 1.95 nW, while the active refresh power of 1.68 nW is comparable, amounting to a total data retention power of 3.63 nW (or 1.7 pW/bit). This total data retention power is comparable to previous reports on low-voltage GC-eDRAM arrays [7], also given for room temperature.

6.5 Macrocell Implementation in 40 nm CMOS

Gain-cell implementations in mature technologies have been frequently demonstrated in the recent past. However, 65 nm CMOS is the most scaled technology in which fabricated GC-eDRAM arrays have been reported to date [15], as discussed in detail in Chap. 2. In this section, we present a 40 nm gain-cell implementation, and explore array sizes and the corresponding minimum operating voltages that result in sufficient array availability.

As previously described, core HVT devices are more efficient than I/O devices for write transistor implementation at scaled nodes, providing similar retention times with relaxed design rules (i.e., reduced area). In addition, the multiple V_T options for core transistors provide an interesting design space for read transistor selection, trading off on- and off-currents, depending on the supply voltage. Two

Table 6.1 Figures of merit for 0.18 μCMOS and 40 nm CMOS ultra-low voltage GC-eDRAM macrocells

Technology node	180 nm CMOS	40 nm LP CMOS
Number of thin metal layers	5	5
Write transistor	PMOS I/O	PMOS HVT
Read transistor	NMOS Core	NMOS SVT
V_{DDmin}	400 mV	600 mV
Storage node capacitance	1.1–4.9 fF	0.27–0.72 fF
Bitcell size	1.12 μm × 3.89 μm (4.35 μm^2)	0.77 μm × 0.42 μm (0.32 μm^2)
Array size	64 × 32 (2 kb)	64 × 32 (2 kb)
Write access time	1 μs	3 ns
Read access time	1 μs	17 ns
Worst-case retention time	40 ms	44 μs
Leakage power	1.95 nW (952 fW/bit)	68.3 nW (33.4 pW/bit)
Average active refresh energy	67 pJ	21.2 pJ
Average active refresh power	1.68 nW (818 fW/bit)	482 nW (235.5 pW/bit)
Average retention power	3.63 nW (1.7 pW/bit)	551 nW (268.9 pW/bit)
Array availability	99.7%	97.1%

additional factors that significantly impact the design at scaled nodes are the reduced storage node capacitance, due to smaller cell area and low-k insulation materials, and severely impeded retention times, due to lower storage capacitance and higher leakage currents. Therefore, array availability becomes a major factor in GC-eDRAM design and V_{DD} selection. For this implementation, a minimum array availability requirement of 97% was defined.

Considering a minimum array size of 1 kb (32 × 32), sufficient array availability is unattainable with the LVT MR implementation for a supply voltage lower than 500 mV, suitable for this device according to Fig. 6.6b. Therefore, an SVT device was considered with near-threshold supply voltages above 500 mV. Figure 6.12a shows the array availability achieved as a function of supply voltage, considering array sizes from 1 to 4 kb. The red dashed line indicates the target availability of 97%, showing that this benchmark can be achieved for a 2 kb array with a 600 mV supply. At this supply voltage, with a −700 mV underdrive write voltage, the write access time is 3 ns, and the worst-case read access time is 17 ns, while the worst-case retention time is 44 μs (see Table 6.1). Figure 6.13 shows the distribution of the time required to sense the discharged voltage of RBL during a read "1" operation following a full retention period (green bars). The red bars (read "0") represent an incorrect readout, caused by a slow RBL discharge through leakage, such that the read access time must be shorter than the first occurrence of an incorrect read "0." The clear separation between the two distributions shows that by setting the read access time to 17 ns, the system will be able to robustly differentiate between the two stored states.

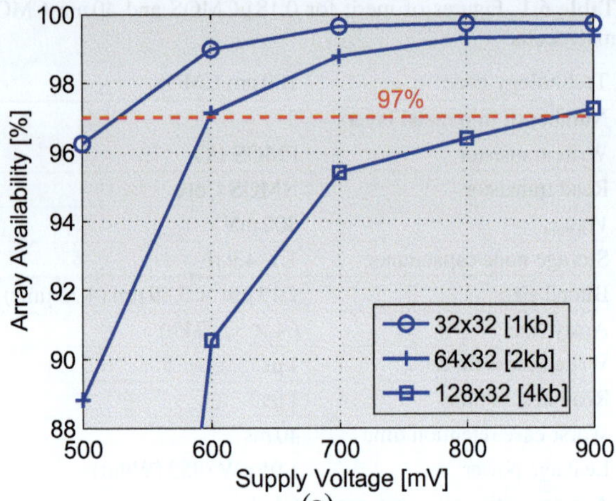

(a)

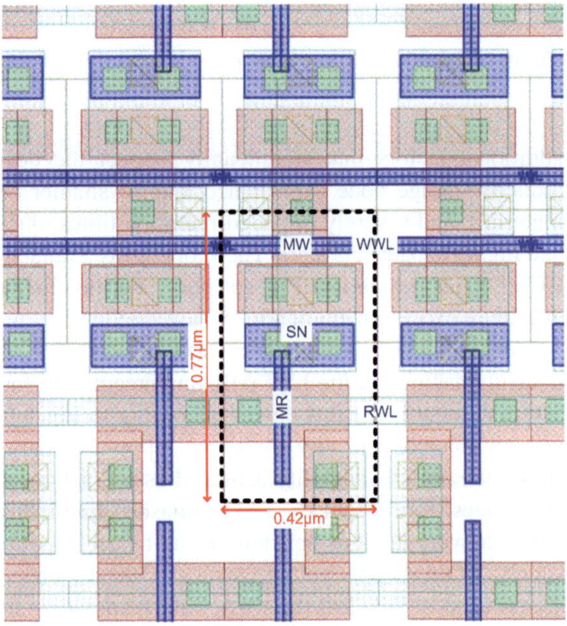

(b)

Fig. 6.12 40 nm gain-cell array: (**a**) array availability as a function of supply voltage and array size; and (**b**) zoomed-in layout

Figure 6.12b shows a zoomed-in layout of the 40 nm gain-cell array with a bitcell area of 0.32 μm^2 (surrounded by the dashed line). For comparison, a single-ported 6T SRAM bitcell in the same node has a slightly larger silicon area of 0.572 μm^2, and robust low-voltage SRAM cells are considerably larger (e.g., the 9T SRAM bitcell in [13] has an area cost of 1.058 μm^2). Table 6.1 show that the implemented 40 nm array exhibits leakage power of 68.3 nW, which is clearly higher than for the array in 0.18 μm CMOS technology. Even though the active energy for refreshing

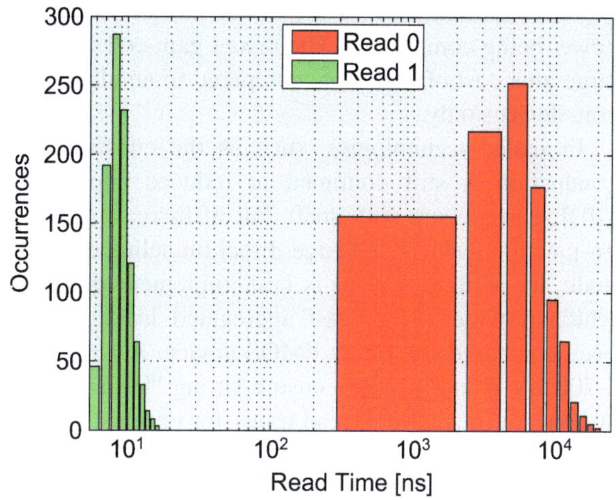

Fig. 6.13 Read access time distribution for the GC-eDRAM implementation in 40 nm CMOS: RBL discharge time for correct data "1" sensing, and undesired RBL discharge time till sensing threshold through leakage for data "0"

the entire array is only 21.2 pJ, the required refresh power of 482 nW is again higher than for the 0.18 µm node, due to the three orders-of-magnitude lower retention time. Consequently, the total data retention power is around 150× higher in 40 nm CMOS, compared to 0.18 µm CMOS.

6.6 Conclusions

This chapter investigated 2-transistor (2T) sub-V_T and near-V_T gain-cell memories for use in ultra-low power systems, implemented in two very different technology generations. First, the retention time model, developed in Chap. 3, was shown to hold for scaled nodes, as low as 28 nm. Next, best-practice approaches for the design of GC-eDRAM bitcells were developed for both mature and scaled technologies.

For mature, above-100 nm CMOS nodes, the main design goals of the bitcell are long retention time and high data integrity. In the considered 0.18 µm CMOS node, a low-leakage I/O PMOS write transistor and an extended storage node capacitance ensure a retention time of at least 40 ms. At low voltages, data integrity is severely threatened by charge injection and capacitive coupling from the read and write word-lines. Therefore, the positive storage node voltage disturb at the culmination of a write operation is counteracted by a negative disturb at the onset of a read operation, which is only possible with a mixed configuration (i.e., an NMOS read transistor). Moreover, the write word-line underdrive voltage must be carefully engineered for proper level transfer at minimum voltage disturb during de-assertion. Monte Carlo simulations of an entire 2 kb memory array, operated at 1 MHz with a 400 mV sub-V_T supply voltage, confirm robust write and read operations under global and local variations. The developed array displays a minimum retention time of 40 ms leading to 99.7% availability for read and write. The total data retention power is estimated to be 3.63 nW/2 kb, the leakage power and the active refresh power being comparable. The mixed gain-cell with a large I/O PMOS device has a large area cost of 4.35 μm^2, compared to an all-PMOS or all-NMOS solution with core devices only.

In scaled technologies, such as the considered 40 nm CMOS node, sub-V_T conduction is still dominant at reduced supply voltages. Gate tunneling and GIDL currents are still small, but of increasing importance, while reverse-biased pn-junction leakage and edge-direct tunneling currents are negligible. In the 40 nm node, the write transistor is best implemented with an HVT core PMOS device, which provides the lowest aggregated leakage current from the storage node, even compared to the I/O PMOS device. A write word-line underdrive voltage of −700 mV is employed to ensure strong "0" levels with a short write access time. Among various NMOS read transistor options, an SVT core device maximizes the sense current ratio between a weak "1" and a weak "0" for near-threshold supply voltages (600–800 mV) where 97% array availability is achieved. Both the access times and the retention time are roughly three orders-of-magnitude shorter than in the 0.18 μm CMOS node, due to the increased leakage currents and smaller storage node capacitance. While the active refresh energy is low (21 pJ), the high refresh frequency results in high refresh power (482 nW), dominating the total data retention power (551 nW). As compared to the 0.18 μm implementation, the scaled down design provides better performance (17 ns read access and 3 ns write access), and a compact bitcell size of 0.32 μm^2.

To conclude, this analysis shows the feasibility of sub-V_T GC-eDRAM operation for mature process technologies and near-V_T operation for a deeply scaled 40 nm process, providing a design methodology for achieving minimum V_{DD} at these two very different nodes.

References

1. Calhoun B, Chandrakasan A (2007) A 256-kb 65-nm sub-threshold SRAM design for ultra-low-voltage operation. IEEE J Solid-State Circuits 42(3):680–688
2. Calhoun BH, Wang A, Chandrakasan A (2005) Modeling and sizing for minimum energy operation in subthreshold circuits. IEEE J Solid-State Circuits 40(9):1778–1786
3. Chiu YW, Lin JY, Tu MH, Jou SJ, Chuang CT (2011) 8T single-ended sub-threshold SRAM with cross-point data-aware write operation. In: Proceedings of IEEE/ACM international symposium on low-power electronics and design (ISLPED), pp 169–174
4. Chun KC, Jain P, Lee JH, Kim C (2011) A 3T gain cell embedded DRAM utilizing preferential boosting for high density and low power on-die caches. IEEE J Solid-State Circuits 46(6):1495–1505
5. Constantin J, Dogan A, Andersson O, Meinerzhagen P, Rodrigues J, Atienza D, Burg A (2012) TamaRISC-CS: An ultra-low-power application-specific processor for compressed sensing. In: Proceedings of IEEE/IFIP international conference on VLSI system-on-chip (VLSI-SoC), pp 159–164
6. Hanson S, Seok M, Lin YS, Foo Z, Kim D, Lee Y, Liu N, Sylvester D, Blaauw D (2009) A low-voltage processor for sensing applications with picowatt standby mode. IEEE J Solid-State Circuits 44(4):1145–1155

7. Lee Y, Chen MT, Park J, Sylvester D, Blaauw D (2010) A 5.42nW/kB retention power logic-compatible embedded DRAM with 2T dual-VT gain cell for low power sensing applications. In: Proceedings of IEEE Asian solid state circuits conference (A-SSCC), pp 1–4
8. Meinerzhagen PA, Andiç O, Treichler J, Burg AP (2011) Design and failure analysis of logic-compatible multilevel gain-cell-based DRAM for fault-tolerant VLSI systems. In: Proceedings of IEEE/ACM great lakes symposium on VLSI (GLSVLSI), pp 343–346
9. Meinerzhagen P, Andersson O, Mohammadi B, Sherazi Y, Burg A, Rodrigues J (2012) A 500 fW/bit 14 fJ/bit-access 4 kb standard-cell based sub-VT memory in 65nm CMOS. In: Proceedings of IEEE European solid-state circuits conference (ESSCIRC), pp 321–324
10. Meinerzhagen P, Teman A, Mordakhay A, Burg A, Fish A (2012) A sub-VT 2T gain-cell memory for biomedical applications. In: Proceedings of IEEE subthreshold microelectronics conference (SubVT), pp 1–3. doi:10.1109/SubVT.2012.6404318
11. Sinangil M, Verma N, Chandrakasan A (2008) A reconfigurable 65nm SRAM achieving voltage scalability from 0.25–1.2V and performance scalability from 20kHz–200MHz. In: Proceedings of IEEE European solid-state circuits conference (ESSCIRC), pp 282–285
12. Sinangil M, Verma N, Chandrakasan A (2009) A reconfigurable 8T ultra-dynamic voltage scalable (U-DVS) SRAM in 65 nm CMOS. IEEE J Solid-State Circuits 44(11):3163–3173
13. Teman A, Pergament L, Cohen O, Fish A (2011) A 250 mV 8 kb 40 nm ultra-low power 9T supply feedback SRAM (SF-SRAM). IEEE J Solid-State Circuits 46(11):2713–2726
14. Teman A, Pergament L, Cohen O, Fish A (2011) A minimum leakage quasi-static RAM bitcell. J Low Power Electron Appl 1(1):204–218
15. Teman A, Meinerzhagen P, Burg A, Fish A (2012) Review and classification of gain cell eDRAM implementations. In: Proceedings of IEEE convention of electrical and electronics engineers in Israel (IEEEI), pp 1–5
16. Teman A, Mordakhay A, Mezhibovsky J, Fish A (2012) A 40-nm sub-threshold 5T SRAM bit cell with improved read and write stability. IEEE Trans Circuits Syst II: Express Briefs 59(12):873–877
17. Teman A, Mordakhay A, Fish A (2013) Functionality and stability analysis of a 400 mV quasi-static RAM (QSRAM) bitcell. ELSEVIER Microelectron J 44(3):236–247

Chapter 7
Novel Bitcells for Scaled CMOS Nodes and Soft Error Tolerance

7.1 Introduction

As shown in previous chapters, GC-eDRAM can bring several improvements over SRAMs in mature process nodes, such as improved density and power consumption. However, aggressively scaled CMOS technologies, characterized by high leakage currents, seriously reduce the data retention time of conventional GC-eDRAM cells due to faster degradation of the storage node voltage. As a result, the refresh power component of these arrays becomes dominant and exceeds the leakage power of SRAMs, making conventional GC-eDRAM cells incompatible with aggressively scaled CMOS nodes.

Moreover, technology and voltage scaling also lead to a rise in the susceptibility of the underlying circuits to radiation-induced soft-errors, often limiting the fabrication of these components to older, more robust process technologies [18, 24]. One of the primary components of application-specific integrated circuits (ASICs) and field-programmable gate arrays (FPGAs) are static random-access memories (SRAMs). The large area of SRAMs makes them highly susceptible to particle strikes, which are common in high radiation environments, such as space. Operation at scaled voltages further reduces the noise margins of SRAMs and significantly increases their susceptibility to radiation effects, such as soft errors caused by single event upsets (SEUs) [5]. In order to guarantee reliable memory operation, an additional power supply must often be incorporated, and complex dynamic voltage scaling techniques are sometimes required [22].

In this chapter, we propose novel GC-eDRAM bitcells for scaled CMOS nodes and soft error tolerance. In Sect. 7.2, we present a novel, 4-transistor GC-eDRAM cell, which uses an internal feedback to reduce the leakage of the storage node, and improve the data retention time of the cell by more than one order of magnitude compared to conventional GC-eDRAM implementations. The cell was integrated into a 1 kB memory array in a 65 nm technology, demonstrating a 10× decrease in retention power as compared to the static power of a 65 nm 6T SRAM.

P. Meinerzhagen et al., *Gain-Cell Embedded DRAMs for Low-Power VLSI Systems-on-Chip*, DOI 10.1007/978-3-319-60402-2_7

In Sect. 7.3, we present a novel solution to embedded memories used in high radiation environments based on a GC-eDRAM, combining both circuit and architectural techniques to provide a soft error tolerant memory, designed for ULP applications. The suggested memory was implemented in a low-power 65 nm CMOS process and showed full compatibility with subthreshold operating voltages of down to 400 mV, with a cell area that is 53% lower than a conventional 6T SRAM and much smaller than alternative rad-hard bitcells, while consuming up to 8× less static power than any other reported rad-hard solutions.

7.2 4T GC with Internal Feedback (IFB) for Scaled CMOS Nodes

While gain cell implementations in mature technology nodes, such as 180 nm, have been shown to display high DRTs of tens to hundreds of milliseconds [21, 27], conventional 2T gain cells in newer technology nodes, such as 65nm, display much lower DRTs of only tens of microseconds [34]. This is a direct consequence of the substantially higher leakage currents which result in a much faster deterioration of the stored levels [27]. Depending on the type of write transistor (WT), one of the data levels has a much higher retention time than the other ("1" for a PMOS WT, "0" for a NMOS WT) [34]. However, when determining the refresh frequency, one must consider the deterioration of the weaker data level under worst-case conditions, i.e., when the write bit-line (WBL) is driven to the opposite level of the stored data during retention periods.

In this section, we present a novel 4-transistor (4T) GC-eDRAM bitcell that selectively protects the weaker data level by means of a feedback loop, thereby decreasing the refresh frequency and reducing the refresh power consumption. Section 7.2.2 presents the implementation and simulation results of a 1 kB memory, based on the proposed 4T GC-eDRAM bitcell. The resulting memory provides a 3× increase in retention time, as compared to the best previously proposed gain cell in the same technology [11], resulting in a 10× decrease in retention power (static plus refresh power) as compared to the static power of a 65 nm 6T SRAM [38]. This is achieved with a 40% smaller bitcell than a 6T SRAM cell in the same technology, allowing for high density, low power integration.

7.2.1 Cell Structure and Operating Mechanism

A standard 2T all-PMOS gain cell [31] is composed of a write transistor (PW), a read transistor (PR), and a storage node (SN), as shown in Fig. 7.1. This cell displays asymmetric retention characteristics with highly advantageous retention of data "1" over data "0." The worst-case condition during retention of a "1" occurs when WBL is grounded and subthreshold (sub-V_T) leakage discharges SN, as illustrated in Fig. 7.1a. However, as the stored level decays to $V_{DD}-\Delta$, the overdrive of PW

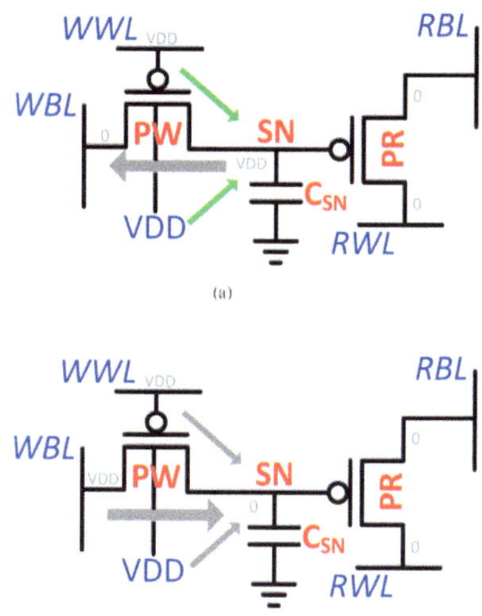

Fig. 7.1 Schematic representations of a conventional 2T gain-cell and its main leakage components: (**a**) Level "1" is stored. (**b**) Level "0" is stored

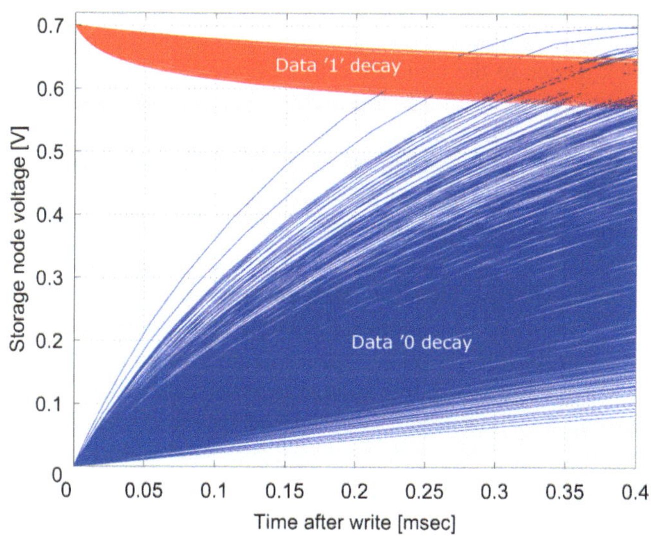

Fig. 7.2 Storage node degradation of a 2T gain cell following a write operation under worst-case WBL bias conditions

($V_{SG} - |V_{Tp}|$) becomes increasingly negative, and the device becomes reverse body biased. Therefore, the sub-V_T leakage is strongly suppressed and the stored level decays very slowly. On the other hand, when a "0" is stored in the cell and WBL is driven to V_{DD}, as illustrated in Fig. 7.1b, this phenomenon does not occur, and the leakage currents gradually charge SN until the data level is lost. These two situations are demonstrated in Fig. 7.2, showing the deterioration of the two data

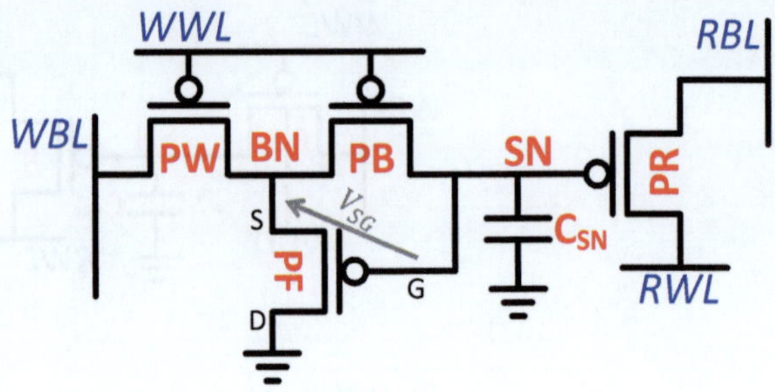

Fig. 7.3 Schematic representation of the proposed 4T gain-cell

levels in a 2T cell, as obtained from 1024 Monte Carlo simulations. This figure, often used to estimate retention time, emphasizes the superiority of the data "1" level in this circuit. It also demonstrates the degraded retention times at scaled technologies, with an estimated DRT of only approximately 200 μs, measured at the earliest intersection between the "0" and "1" samples. Note that this is only a rough estimation of DRT, since for a full DRT evaluation, the array architecture and the read scheme need to be taken into account [34].

The immediate conclusion from the phenomena presented above is that the data "0" DRT is the bottleneck that needs to be resolved in order to increase the retention time of this bitcell. The proposed cell addresses this by adding a buffer node (BN) and a feedback device to the basic 2T cell, as shown in Fig. 7.3. SN is connected in a feedback loop to the feedback device (PF), which conditionally discharges the BN according to the stored data state. An additional buffer device (PB) separates the stored data level from the BN to ensure extended retention time. The resulting 4T bitcell is composed of standard threshold-voltage (V_T) transistors and is fully compatible with standard CMOS processes. PMOS devices are selected over NMOS due to their lower sub-V_T and gate leakages to provide longer retention times while consuming a small cell area. Detailed cell operation is explained hereafter.

Cell access is achieved in a similar fashion as with a standard 2T cell. During writes, the write word-line (WWL), which is connected to the gates of both PW and PB, is discharged to a negative voltage in order to enable a full discharge of SN (when writing a "0"). Readout is performed by pre-discharging the read bit-line (RBL) to ground and subsequently charging the read word-line (RWL) to V_{DD}. RBL is then conditionally charged if the storage node holds "0," and otherwise remains discharged. To save area and power, a simple sense inverter is used on the readout path; however, conventional sense amplifiers can be used for improved read performance.

The novelty of the proposed cell becomes apparent during standby periods, when the internal feedback mechanisms come into play. During hold, PW and PB are off (WWL = V_{DD}), and we assume worst-case retention conditions, i.e., that WBL is driven to the opposite voltage of the stored data level. For a stored "1," a self-limitation mechanism, similar to that of the standard 2T cell, ensures that the level decays slowly. In addition, the transistor stack (PW and PB) provides more resistance between SN and WBL and results in even lower leakage and a slower decay compared to the standard 2T cell. For data "1," PF is in deep cutoff, such that its effect on the circuit is almost negligible. However, following a write "0" operation, V_{SG} of PF is equal to the voltage at BN (V_{BN}). This is much higher than the negative V_{SG} of PB, and therefore any charge that leaks through PW to BN will be discharged through PF and not degrade the "0" level at SN. In this way, the worst-case condition of the 2T cell is avoided and retention time is significantly increased. In summary, the feedback path protects the weak "0" state on the SN by pulling BN to ground, while the worst-case V_{DD} drop across PW and the corresponding sub-V_T leakage do not affect the retention time of the cell, as the feedback path is disabled for the strong "1" level. Note that the proposed technique only delays the decay of a "0" level, but cannot fully avoid it: gate tunneling through PR, as well as the GIDL and junction leakage of PB still charge SN, while sub-V_T leakage of the turned-off PB counteracts (but does not avoid) this SN charging process.

7.2.2 Implementation and Simulation Results

A 64×32 bit (2 kb) memory macro based on the proposed cell was designed in a low-power CMOS 65 nm process. All devices were implemented with standard V_T transistors to provide full compatibility with a logic process. The operating voltage was selected to be 700 mV, to demonstrate compatibility with power-aware (near-threshold) applications. The circuit layout, shown in Fig. 7.4, was carried out according to standard design rules, resulting in an area of 0.708 μm^2 (0.92 $\mu m \times 0.77$ μm), which is 40% smaller than a reference 6T SRAM bitcell in this technology (laid out using standard design rules).

Cell operation is demonstrated in Fig. 7.5 through subsequent write and read operations to the proposed 4T gain cell. Initially, a "0" is written to SN by pulsing WWL to a negative voltage (−700 mV), thereby discharging SN through WBL. Next, a read operation is performed by pre-discharging RBL by pulsing the PC signal, and subsequently charging RWL. As required, RBL is driven high through PR. Prior to the next assertion of WWL, WBL is driven high in order to write a "1" to SN. During the next read cycle, the pre-discharged RBL remains low, as the stored "1" level cuts-off the discharge path through PR.

DRT estimation plots for the 4T cell are presented in Fig. 7.6 for comparison with those presented in Fig. 7.2. Again, 1024 MC samples were simulated in a 65 nm CMOS process with a 700 mV supply, driving WBL to the opposite voltage of that stored on SN. The level degradation of Fig. 7.6 is not only much more balanced than the asymmetric degradation of the 2T cell, but it is also more than an order of magnitude higher. The estimated DRT, extracted from these plots, is 8.29 ms at 27 °C and 3.98 ms at 85 °C. This is over 3× higher than the best retention time reported so far in a 65 nm CMOS node [13]. Moreover, the symmetric behavior of the two data states is more appropriate for differentiating between "0" and "1"

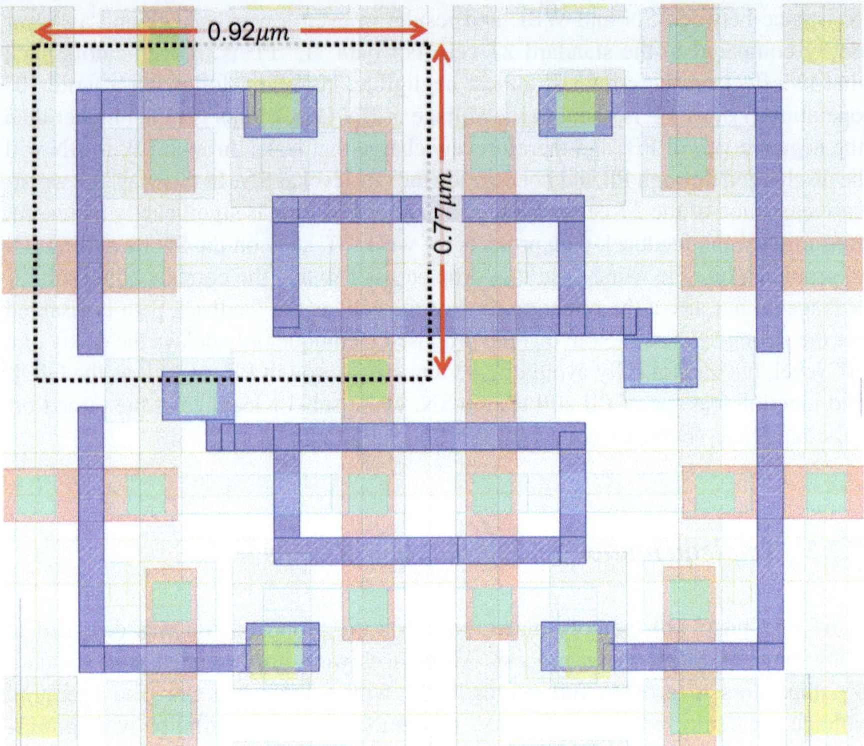

Fig. 7.4 4T Bit-cell layout

levels, easing the design of the readout circuit and potentially further enhancing the actual retention time (latest successful read) compared to the standard 2T cell.

Chun et al. [13] previously showed that a standard 2T GC-eDRAM can consume lower retention power than a similarly sized SRAM in 65 nm CMOS. Since the retention time of the presented gain cell, as shown previously, is over 40× higher than that of a standard 2T cell, the retention power, composed of leakage and refresh power, is even lower. For the proposed 4T-bitcell memory macro, the retention power was found to be 3.86 pW/bit at 27 °C and 53.78 pW/bit at 85 °C, which is almost 5× less than the leakage power of a 6T-bitcell SRAM operated at a 0.7 V supply voltage. A comparison between the proposed cell and other embedded memories is given in Table 7.1. The table clearly emphasizes the benefits of this cell, achieving much lower retention power due to its increased retention time.

Performance of the proposed 4T cell is summarized in Table 7.2. At 700 mV, the active refresh energy is 6.89 fJ/bit, composed of 5.88 fJ/bit for read and 1.01 fJ/bit for write. The cell has a read delay of 2.32 ns (using a slow but small sense inverter) and a write delay of 0.4 ns (with and underdrive of −700 mV). A standard 2T gain-cell was measured to have a 0.29 ns write delay, which is the same order of magnitude as the proposed cell.

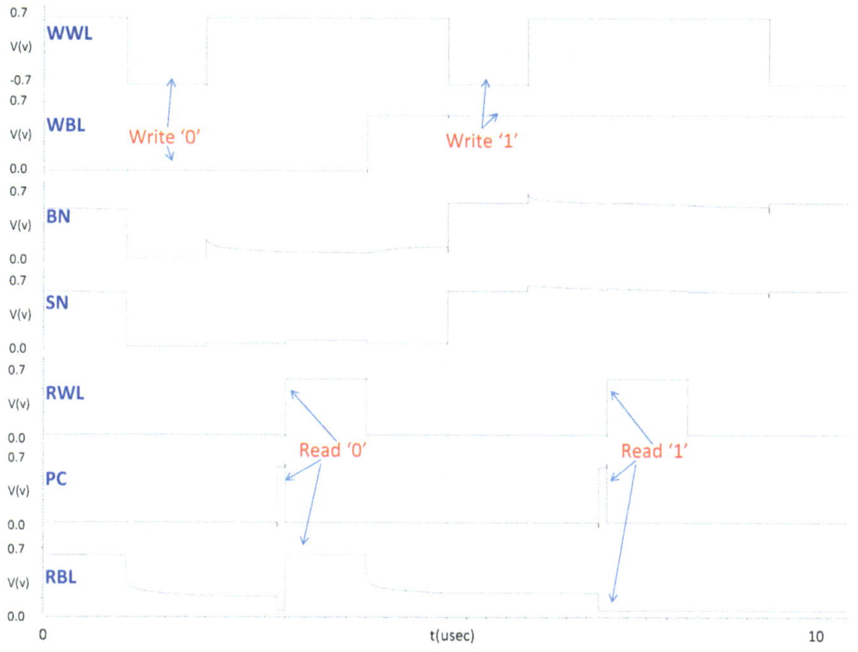

Fig. 7.5 Timing diagrams demonstrating circuit operation

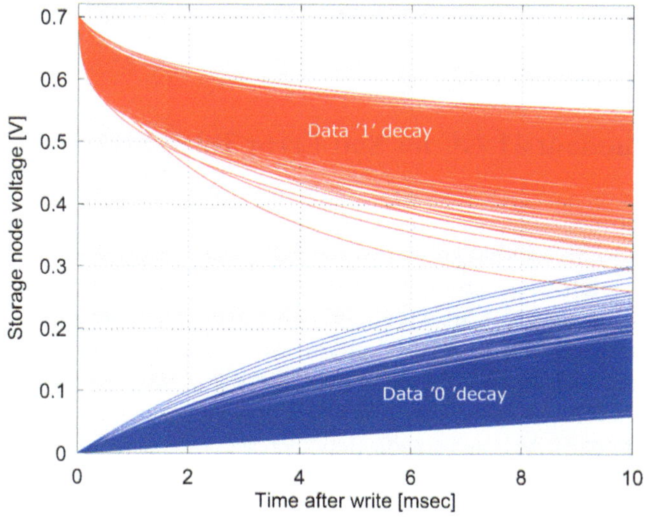

Fig. 7.6 Storage node degradation of a 4T gain cell following a write operation under worst-case WBL bias conditions

Table 7.1 Comparison between proposed design and other embedded memory options

	6T SRAM [38]	2T1C gain cell [14]	2T gain cell [31]	Proposed 4T gain cell
CellStructure	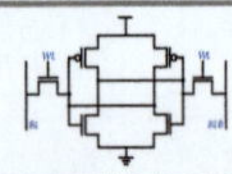	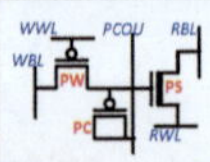	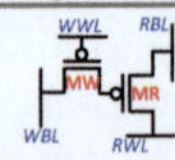	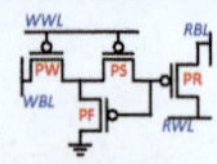
DrawnCellSize	1.18 μm^2(1X)	0.69 μm^2(0.58X)	0.27 μm^2(0.23X)	0.71 μm^2(0.6X)
SupplyVoltage	1.1 V	1.1 V	1.1 V	0.7 V
RetentionTime	Static	0.5 ms	10 μs	3.98 ms
RetentionPower	264.58 pW@0.7 V 564.29 pW@1.1 V	158 pW	1.95 μW	53.78 pW@0.7 V 126.9 pW@1.1 V

All designs are in 65nm CMOS.
Simulation were conducted at @85°C.

Table 7.2 4T gain cell performance summary

Technology	65 nm LP CMOS
Cell area	0.708 μm^2
4T eDRAM / 6T SRAM cell area ratio	0.6
Supply voltage	700 mV
Worst-case retention time	8.29 ms@27 °C
	3.98 ms@85 °C
Write delay (worst)	0.4 ns@85 °C
Read delay (worst)	2.32 ns@85 °C
Active read energy	5.88 fJ/bit@85 °C
Active write energy	1.01 fJ/bit@85 °C
Active refresh energy	6.89 fJ/bit@85 °C
Leakage power/bit	2.87 pW@27 °C
	51.29 pW@85 °C
Retention power/bit	3.86 pW@27 °C
	53.78 pW@85 °C

7.3 Redundant 4T GC for Soft Error Tolerance

One of the primary components of ASICs and FPGAs are SRAMs, and the amount of SRAM in VLSI SoCs has grown exponentially over the past few decades. SRAMs are nowadays often responsible for over 50% of the total area and static power consumption of modern ASICs [19]. This dominant area share of SRAMs makes them highly susceptible to particle strikes, which are common in high radiation environments, such as space. Furthermore, modern ASICs are often operated at scaled supply voltages in order to reduce their power consumption. This reduces the noise margins of SRAMs and significantly increases their susceptibility to radiation effects, such as soft errors caused by SEUs [5]. In order to guarantee reliable memory operation, additional power supplies are often incorporated, and complex dynamic voltage scaling techniques are sometimes required [22].

Embedded memory errors are typically handled at either the architectural or circuit level [2]. At the architectural level, redundancy schemes, such as error correction codes (ECCs) [4, 23], dual modular redundancy (DMR) [33] and triple modular redundancy (TMR) [32] are commonly used. However, these solutions are extremely costly in terms of area overhead, while also resulting in high complexity and performance degradation. Therefore, they are not suitable for small, high-speed cache memories such as level-1 (L1) caches. Mohr et al. [28] suggested using two-dimensional parity checking and bit interleaving techniques as a solution for this problem. However, conventional architectural solutions do not provide sufficient protection under voltage scaling, which is a common approach to low-power systems.

Circuit level techniques such as DICE [8] and Quatro 10T [20] can efficiently increase SEU tolerance. These solutions achieve SEU tolerance by increasing the internal feedback of the static latch core of a conventional 6T bitcell. This approach increases the critical charge (Q_{crit}) of the bitcell; however, once this value is exceeded, the positive feedback is actually what ensures a bit-flip. Furthermore, these cells feature a much higher transistor-count than traditional SRAM, resulting in up to 2× increase in area [10] and power consumption [20] of the memory array. Therefore, these cells are inefficient for high-speed applications, such as register files and caches that demand full hardening and high density [14, 36].

In this section we take a completely different and nonintuitive approach to circuit-level SEU tolerance by entirely removing this feedback from the cell. We propose to use a dynamic memory core based on GC-eDRAM, which by nature is more susceptible to soft-errors than a static memory cell. However, the reduced transistor count and the physical properties of the dynamic circuit allow us to internally apply complementary dual-modular redundancy (CDMR) to achieve inherent per-bit error detection. In addition, the simple addition of parity provides error correction capabilities at a much lower overhead than traditional ECC-based approaches.

The proposed CDMR memory was implemented in a low-power 65 nm CMOS process, consuming less area than any other previously proposed soft-error tolerant memory cell, and showing full compatibility with voltage scaling down to 400 mV. The resulting cell area is 53% smaller than a conventional non-radiation hardened 6T SRAM bitcell, and consumes up to 8× less static power than any other reported rad-hard solution.

7.3.1 Radiation-Hardened Memories

An SEU is a change of state in a flip-flop or memory cell caused by charged particles striking a sensitive node in an IC device, potentially causing system failure. SEUs occur when an energetic particle passes through a silicon substrate and its energy is transferred into the creation of electron-hole pairs along its path [2]. When such a particle hits a reversed-biased junction inside a storage node of a memory circuit, the resulting current pulse can inject enough charge into the junction to cause a data flip [6, 15].

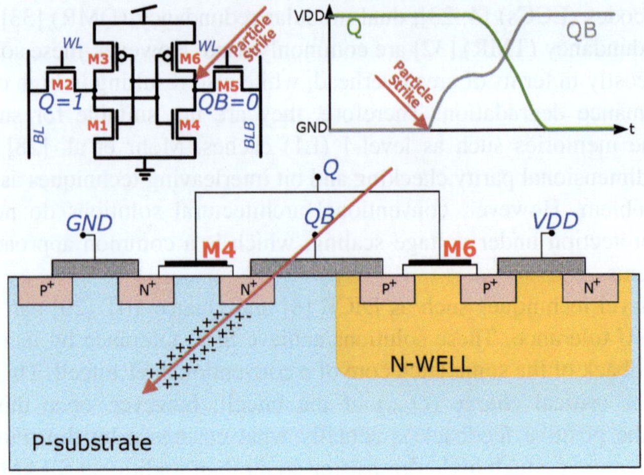

Fig. 7.7 Particle strike hitting a silicon substrate junction, releasing electron-hole pairs, and causing a state flip in an SRAM cell

The conventional 6-transistor (6T) SRAM memory cell, shown in the inset of Fig. 7.7, utilizes an active feedback loop between cross-coupled inverters (M1/M3 and M4/M6) in order to retain its stored data value. This circuit is very sensitive to SEUs, as any upset that causes one of the data nodes to cross the switching threshold of the adjacent inverter will result in a data flip. An example of such an occurrence is illustrated in Fig. 7.7, showing a particle striking the drain of one of the NMOS pull-down devices of a 6T SRAM bitcell (M4, in the inset). If the particle injects sufficient charge to drive the voltage at the QB node past the switching threshold of the opposite inverter (M1/M3), the feedback loop will drive the storage state of the SRAM cell to the opposite level, as shown in the illustrative waveforms.

The failure risk, described above, increases with process scaling, since the critical charge of the memory cell decreases, resulting in higher soft error rates (SERs). Furthermore, static noise margins are decreased significantly [7], and therefore the memory cell is more susceptible to read and write errors. In addition, when operating at low voltages for power reduction, the aforementioned switching threshold decreases, thereby increasing the error susceptibility of the circuit. These issues have led to the development of SEU-aware SRAM design approaches for radiation-hardening.

The main approach to error mitigation over the past few decades has been to provide full immunity to errors through circuit redundancy. By simply adding a parity bit to a given number of bits, a single error can be detected with very little overhead. For enhanced protection, ECCs can be used to detect and correct multiple errors. However, the error detection and correction capabilities of such codes are directly correlated with the complexity and area overheads required for their implementation. For complete protection under the assumption that the probability of an error at two separate places on the chip within a defined timespan is extremely low, triple modular redundancy replicates each storage node three times and uses a majority gate to determine the correct value. While TMR is a simple and robust solution, it incurs a significant increase in both area and power of over 3×.

The alternative approach to providing error-tolerance is to design circuits that are directly aimed at improving the soft error susceptibility of the memory array. Recently proposed error-tolerant SRAM bitcells include the temporal latch [25], DICE [8], the Quatro-10T and 12T bitcells [20, 30], the 13T subthreshold bitcell [1], and SHIELD [10]. The primary goal of these circuit solutions is to increase the Q_{crit} of the cell, and thereby improve their resiliency to SEUs. These solutions can be fabricated in commercially available, state-of-the-art manufacturing processes and can be combined with the aforementioned architectural approaches to provide enhanced error detection and correction capabilities. However, all previously proposed radiation-hardened bitcells are much larger than standard SRAM and often require additional peripheral circuits and signals for correct operation. In addition, their high transistor count leads to their increased static power consumption. Therefore, while these circuits can provide a viable solution for radiation-hardened storage, this comes at the expense of large area and power overheads, which contradict the aforementioned limitations of space-bound systems.

7.3.2 Proposed 4T CDMR Dynamic Memory Array

The primary mechanism that enables an SEU to flip the data in a conventional SRAM cell is the positive feedback between the two internal storage nodes. While this feedback is the means by which the static storage capabilities of the circuit are acquired, in the case of SERs, it is also the cell's vulnerability. Any voltage shift which causes one of the storage nodes to cross the switching threshold of the adjacent inverter will immediately be latched by the positive feedback, resulting in a catastrophic bit flip. Such a voltage shift can be quantified according to the charge transferred by the striking particle, and therefore, the critical charge metric is used to characterize the susceptibility of circuit-level solutions to SEUs. If a particle strike induces charge lower than the Q_{crit} of the circuit, the storage value will remain intact; however, if Q_{crit} is exceeded, a failure will occur.

While SRAM is the primary technology used to implement embedded memory arrays, another popular storage topology is embedded dynamic random access memory (eDRAM). GC-eDRAM is a fully logic-compatible implementation of eDRAM, which provides a reduced silicon footprint as compared to SRAM, but lacks the internal feedback that ensures strong storage levels, in spite of deteriorating leakage currents [35]. Intuitively, such a topology is much more susceptible to SERs, as the circuit lacks any mechanism to mitigate a level change induced by a particle strike. However, we propose to employ complementary dual-modular redundancy, i.e., for each bit, to store both the data value and its inverse. Based on this concept, SEUs in the memory array can be both detected and corrected, while still achieving the low area and power aspirations of the target applications.

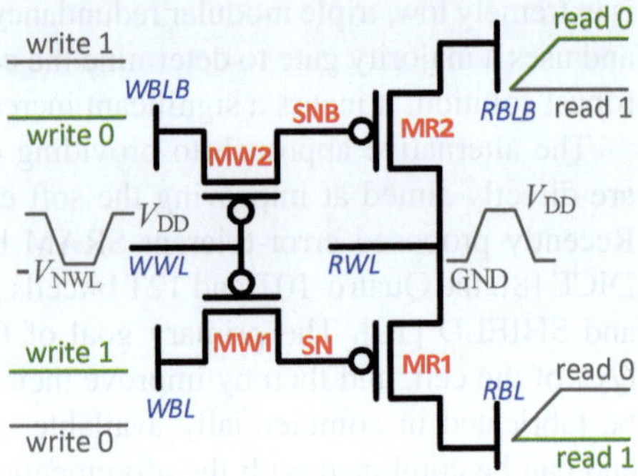

Fig. 7.8 Schematic representations of the proposed 4T gain cell

The proposed CDMR approach is implemented with the four-transistor (4T) dynamic memory bitcell shown in Fig. 7.8. The circuit consists of two write transistors (MW1 and MW2), two read transistor (MR1 and MR2), and two storage nodes (SN and SNB). The data and its complementary values are stored on the parasitic capacitances of the storage nodes, comprising the gate capacitance of MR1/MR2 and the diffusion capacitance of MW1/MW2, respectively. Writing to the cell is achieved by driving the write word line (WWL) to a negative voltage and passing the data and its complementary level from the write bit lines (WBL and WBLB) to SN and SNB, respectively. Readout is achieved by pre-discharging the read bit lines (RBL and RBLB) to GND and driving the read word line (RWL) to V_{DD}, thereby charging RBL/RBLB only if SN/SNB holds a data "0." The RBL/RBLB of each column is connected to a sensing circuit, which is implemented with a simple inverter, in order to output the digital levels of the data and its complementary level stored in the selected cell.

During normal operation, the output of a read operation from a single bitcell will provide two opposite levels. If an SRAM cell is exposed to a particle strike surpassing the Q_{crit} of one of the storage nodes, the feedback will cause the complementary node to flip, as well as latching the erroneous value. However, since the proposed memory structure lacks a similar internal feedback mechanism, the complementary storage nodes are affected separately by a particle strike. In the case of a particle changing the data in one of the storage nodes, the complementary node will remain unaffected and both storage nodes will store the same data level.

Moreover, due to the choice of an all-PMOS implementation of the proposed bitcell, only a "0" to "1" upset can occur, resulting in both the data (SN) and its complementary value (SNB) storing a "1" following such an upset. Therefore, by integrating a single AND gate on the readout path of every column of the array, the outputs are compared and an error can be detected in any given bit.

In itself, the CDMR characteristic of the proposed topology provides a reduced area bitcell implementation with inherent multiple-bit error detection capabilities. However, while the location of the error is known, it is not inherently clear if the data or its complementary value was corrupted, as both of the values will store a logic "1." This prevents the straightforward correction of the error. However, error-correction capability can be achieved by simply adding a parity bit to each set of bits.

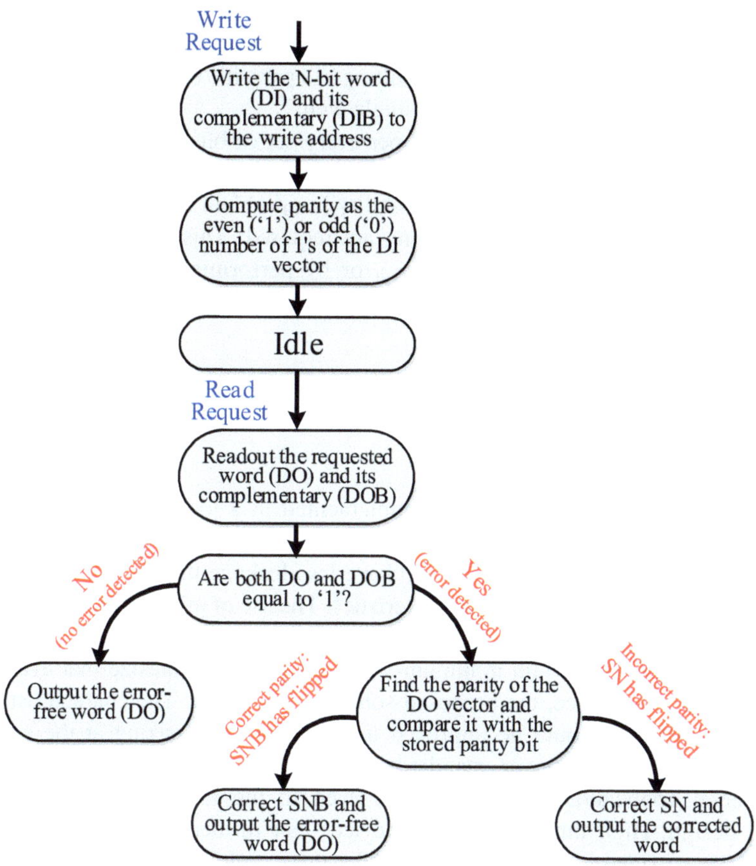

Fig. 7.9 Flowchart illustrating the error detection and correction algorithm

The parity bit for every set of N bits is written during write access to the memory and indicates whether the number of ones in the written set of bits is even or odd. If the parity is maintained despite the error, the complementary value (SNB in Fig. 7.8) of the erroneous bit has been corrupted and should be corrected, while if the parity is incorrect, the erroneous SN value should be corrected.

A flowchart demonstrating the error detection and correction algorithm is demonstrated in Fig. 7.9. First, an N-bit word (DI) and its complementary (DIB) are written to the input write address. In parallel, the parity of DI is computed and stored in the memory array. Following some period of retention, the requested word in the array is read, outputting the data (DO) and its complementary value (DOB). These two values are compared bit-wise, and if any of the bits are equivalent, an error has occurred, and a correction mechanism checks the parity of the DO vector against the stored parity bit. If the parity is found to be correct, the SNB of the erroneous bit is incorrect, while if the parity is incorrect, the SN value has flipped. The erroneous data is then corrected to provide an error-free output.

Note that the suggested error correction scheme with a single parity bit for an N-bit word does not apply to multiple bit upsets, which can be induced through highly energetic cosmic-ray particles resulting in a parasitic bipolar conduction [29], potentially flipping numerous bits in a single well.

To account for more than one error per every set of bits that includes a parity, bit interleaving techniques can be used [23]. However, for single-ended readout storage architectures, half-select susceptibility is a problem that must be addressed, if column multiplexing or byte masking are required. This can easily be solved by either writing entire words at a time, or by performing a read-modify-write operation, when byte masking is required. Moreover, parity can be applied to every $k < N$ bits for additional protection.

7.3.3 *Implementation*

The 4T dynamic memory array was implemented in a low power 65 nm CMOS technology using standard V_T and minimum-sized PMOS transistors. PMOS transistors were selected since they tend to have less leakage than NMOS transistors, resulting in better retention time characteristics. The use of minimum-length devices is not optimal in terms of retention time and/or static power. However, we chose to demonstrate the idea with minimum-sized devices to emphasize that even with this minimal area choice, the proposed solution is efficient. The characteristics of the proposed bitcell could be improved through device up-sizing at the expense of additional area. Moreover, standard V_T devices were chosen to demonstrate the general application of the approach even to the most basic and compact circuit design. High threshold (HVT) devices can further increase the retention time, at the expense of an increased under-drive voltage for write assist. The combination of HVT and standard-V_T devices for implementation of the write and read ports, respectively, can be used to provide better retention time and read performance. However, this too comes at the expense of area required by the design rules for mixing these V_T options. For further discussion of these and other GC-eDRAM implementation options, the reader is encouraged to turn to previous publications [17, 27].

The following subsections demonstrate the bitcell's SEU tolerance, layout, and power comparison with a conventional 6T SRAM bitcell and the state-of-the-art radiation-hardened circuit-level solutions.

7.3.3.1 SEU Tolerance

Cell operation under SEUs is demonstrated in Fig. 7.10, showing consecutive write, upset, and read events for data "1" and "0" with a 400 mV supply voltage, which is suitable for low-power applications, often operating in the subthreshold domain. The SEU was modeled by connecting a current source to the SN node of the bitcell and applying a double-exponential current pulse during the hold states of the cell,

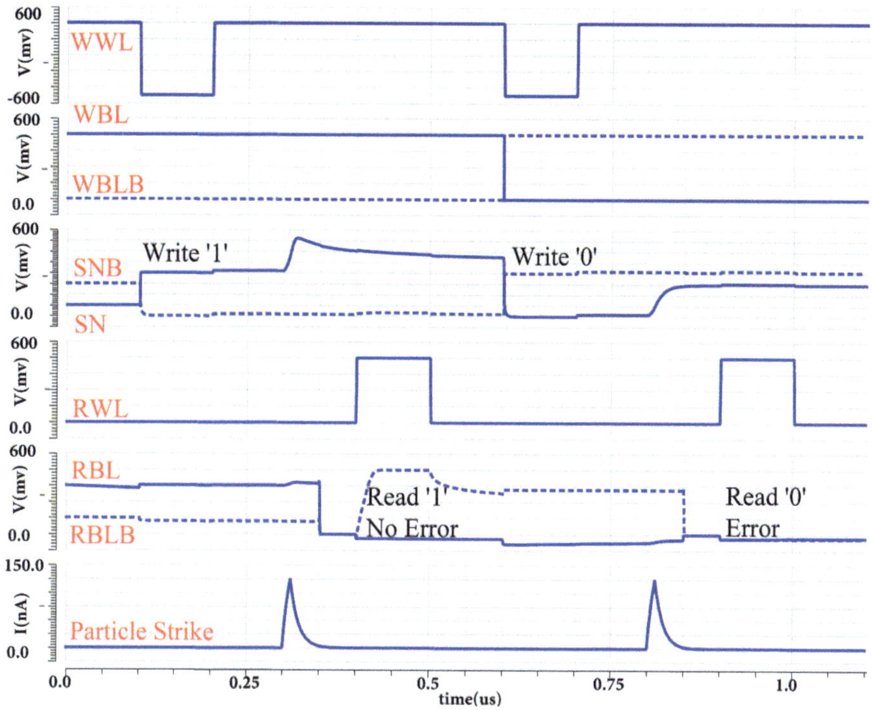

Fig. 7.10 Waveform demonstration of write-upset-read events

according to the model presented in [16]. The physical composition of the cell only allows a positively charged upset [1], and therefore, in the example, only this type of particle strike is shown. On the one hand, in the first demonstrated strike, the cell is storing a "1," and the positive charge only strengthens the stored level, leaving RBL discharged during readout and not leading to an error. On the other hand, when data "0" is stored, the applied pulse causes an increase in the voltage level stored at SN, causing the RBL to erroneously charge during readout. Since the complementary value, stored in SNB, is also a data "1," both RBL and RBLB provide a logic "1" at the output, indicating that an error has occurred. In addition, the parity will no longer be correct, which implies that the error occurred in the SN of the erroneous bit, and therefore, this node should be discharged back down to "0" to correct the error.

As pointed out previously, the sensitivity of a memory bitcell to particle strikes is commonly quantified using the Q_{crit} metric, which is determined by integrating the applied current source at the point where it causes a read failure following an SEU. However, the inherent error detection and correction capabilities of the proposed topology and architecture significantly reduce the relevance of the Q_{crit} metric and make it inappropriate for comparison with other solutions. Nonetheless, the Q_{crit} of the 4T bitcell was extracted through simulation under particle strikes, according to [16]. Interestingly, due to the dynamic nature of the cell and the deterioration of

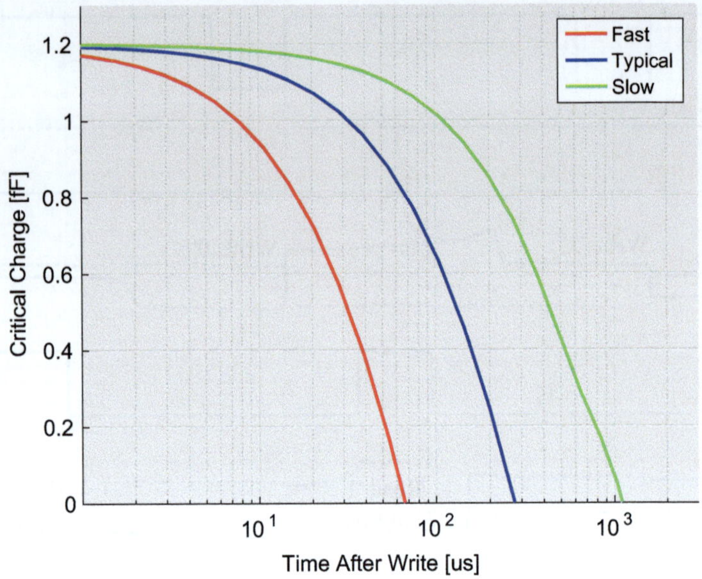

Fig. 7.11 Q_{crit} as a function of the time after write at different process corners

the stored level between refresh operations, the Q_{crit} varies with time. The results are shown in Fig. 7.11 for a bitcell under worst-case retention time conditions [27] in different process corners, showing that the Q_{crit} has the highest value following a write "0" operation and it degrades as time passes due to a deteriorated "0" level as a result of leakage. For a fast corner the Q_{crit} is equal to 0 after 50 µs, since the retention time has passed, and any particle strike would cause an error. For typical and slow corners, the Q_{crit} equals 0 after 340 µs and 1.7ms, respectively. While these values are significantly smaller than those presented by other circuit-level solutions, the error-correction and detection capabilities of the cell make them sufficient for high-radiation operation.

With aging, it is assumed that the critical charge of the cell will increase, as aging usually affects CMOS processes by causing the threshold voltage to rise due to hot carriers and negative bias temperature instability (NBTI). Such effects actually improve the retention time of GC-eDRAM cells, as the subthreshold leakage from the storage node decreases. Moreover, previous publications [3, 9] have shown that NBTI degradation does not have a large impact on the single-event upset rate.

7.3.3.2 Silicon Footprint

The standard approach to memory bitcell design is to use very small devices and "push" the design rules to their limits in order to integrate as many bits as possible within a given area. While the resulting noise margins and soft-error tolerance of 6T and 8T cells designed with pushed rules are sufficient for standard operating

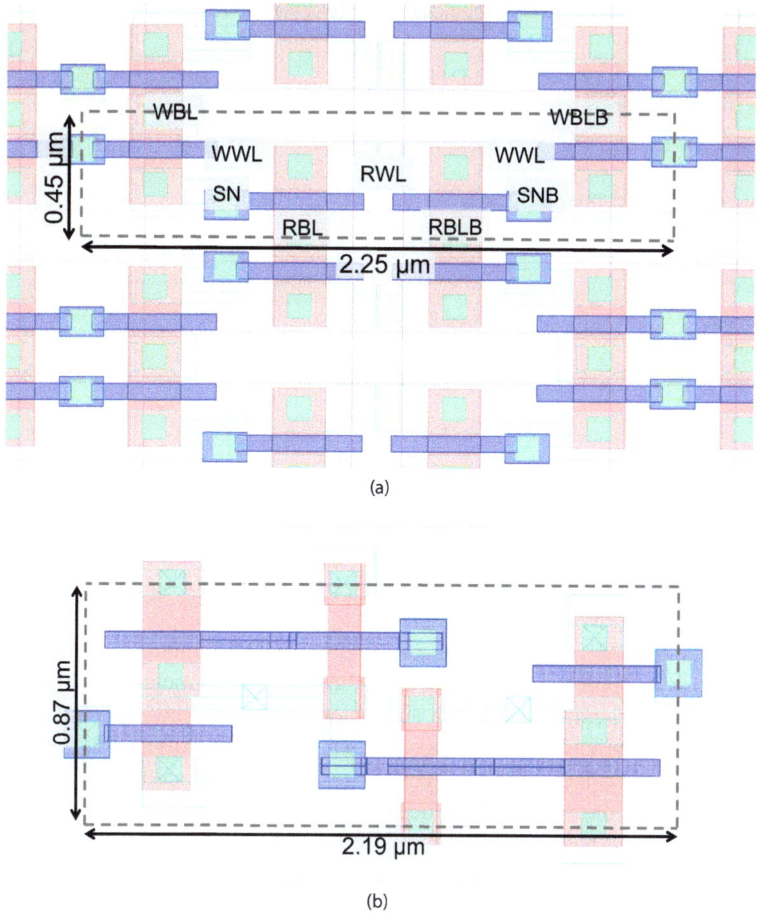

Fig. 7.12 (**a**) Small section of the 4T GC-eDRAM array layout showing the dimensions of the unit cell compared to (**b**) 6T SRAM cell layout

environments, this approach leads to circuits that are extremely sensitive to high-radiation environments. Therefore, all previous radiation-hardened bitcells, such as DICE [8], Quatro 10T [20], and SHIELD [10], incur a large area overhead, as compared to a conventional, high-density 6T bitcell. The proposed approach, on the other hand, reduces both the transistor count and the overall silicon footprint of the memory bitcell, while still providing SEU tolerance, as described earlier.

The layout of the proposed 65 nm 4T CDMR bitcell, implemented with standard-V_T PMOS transistors, is shown in Fig. 7.12a. The figure presents a single bitcell (marked with a dashed line), integrated in a memory array, comprising an area of 1.01 µm^2. For comparison, a conventional 6T bitcell, drawn with standard design rules, is illustrated in Fig. 7.12b. The proposed cell is 47% smaller than the standard

Table 7.3 Comparison of memory solutions for space applications

Memory type	6T SRAM	SHIELD [10]	12T DICE [8]	Proposed 4T CDMR memory
Cell size	451 f^2	947 f^2	861 f^2	240 f^2
Cell static power @ 1.2 V	54.9 pW	35.7 pW	58.3 pW	14.7 pW
Cell write energy @ 1.2 V	10.78 fJ	12.26 fJ	15.39 fJ	6.28 fJ
Cell read energy @ 1.2 V	9.1 fJ	11.89 fJ	14.21 fJ	5.87 fJ
Protection mechanism	–	Increased Q_{crit}	DMR	CDMR
Error detection	Requires parity or ECC for each detection		–	Inherent per-bit detection
Error correction	Requires ECC (e.g., 7 bits for 1 error in 32-bit word)			1-bit parity per error

The reported 6T SRAM cell size with pushed layout rules is 131 f^2

non-SEU tolerant memory circuit, and between 2.5× and 5× smaller than the other considered bitcells, as reported in Table 7.3.

Note that SRAM macros are often provided with high-density "pushed-rule" layouts. Such an approach is, in most cases, unavailable to the designers of custom bitcells. Therefore, comparison with an implementation following standard design rules is common practice in literature [12, 13, 31]. In addition, such an approach presumably further increases the SEU sensitivity of the cell, and is therefore detrimental in the case of rad-hard design. Nonetheless, for further reference, the reported area of pushed-rules SRAM in the same technology node is given in the footnote of Table 7.3.

7.3.3.3 Power Consumption

The static power of memory arrays is often considered the most important aspect of chip power consumption, due to the large number of bitcells in a standby state at any given time. While leakage is the dominant power component of a standard SRAM, the total retention power of dynamic memories comprises both the leakage and the refresh power of the array, and in most cases, the refresh power is the dominant factor [27]. Since the refresh power is a function of the DRT, extraction of this parameter is essential for retention power estimation.

Figure 7.13 presents the DRT distribution of the proposed bitcell, extracted for a worst-case scenario, as described in [27, 35] from 8 k Monte Carlo statistical simulations, modeling both process variations and local variations (mismatch). Based on this data sample, the worst-case DRT for this technology would be 50 µs. Of course, for large memories, a higher sigma DRT estimation is needed. Therefore, a power transformation was applied to the distribution of Fig. 7.13 to extrapolate

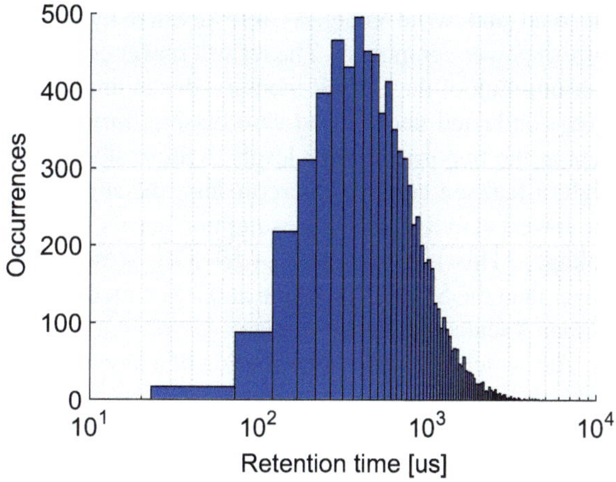

Fig. 7.13 DRT distribution of proposed bitcell

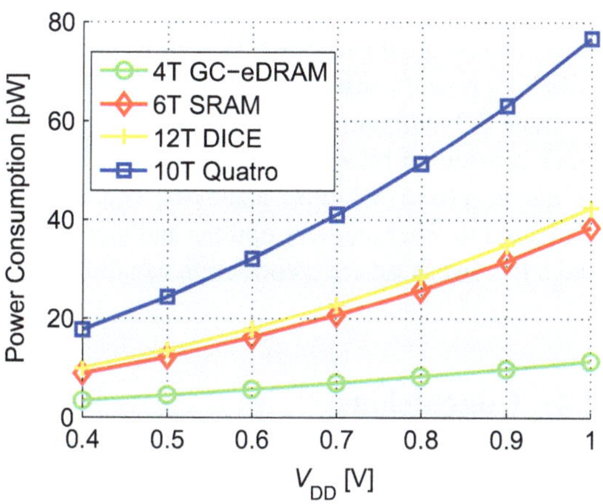

Fig. 7.14 Comparison between the retention power of the proposed 4T gain cell and the leakage power of a 6T SRAM, 12T DICE [8], 10T Quatro [20], and SHIELD [10]

the six-sigma worst-case DRT [37], which was found to be 12 µs. However, when considering large memories, constructed of many banks and arrays, separate refresh periods can be applied to each array. Furthermore, refresh power can be further reduced through various retention-time extension and process compensation techniques, such as those shown in [26, 35]. Therefore, for our power calculations, we consider the DRT extracted from the statistical sample of Fig. 7.13, as a typical refresh period value for an average GC-eDRAM bank.

Figure 7.14 shows a comparison between the retention power of the suggested cell with the static power consumption of a conventional 6T SRAM, 12T DICE [8], 10T Quatro [20], and SHIELD [10] bitcells, which were implemented in the same process technology with minimal sized transistors. The retention power constitutes of the leakage power and refresh power of the cell, which includes the switching power of read and write operations. Therefore, the first step was to evaluate the retention time of the cell to indicate the required refresh period. Subsequently, the read and write energies were divided by the retention time to provide the refresh power component. The refresh power constituted 40–60% of the total power consumption of the CDMR memory, depending on the supply voltage. Simulations were conducted under worst-case biasing during retention with WBL and WBLB kept at the opposite voltage levels to those stored at SN and SNB, resulting in the highest leakage currents. Nevertheless, the suggested CDMR memory clearly has the lowest standby power consumption across the entire range of simulated supply voltages. This is a reduction of 59–87% at nominal conditions (1.2 V) and 48% lower than the SHIELD solution at 0.4 V, which is the only other functional solution at such a scaled operating voltage.

The switching power consumption of a memory array is proportional to its bitline and wordline capacitances, which need to be charged/discharged during a dynamic operation. The smaller size of the suggested cell, as compared to the other bitcell solutions, indicates that the switching power of the suggested cell is also lower due to the smaller capacitances that need to be charged. The dynamic energy required to write and read a single bitcell was simulated using layout extracted parasitic capacitances and the results clearly indicate that the suggested 4T cell has the lowest switching power consumption of the considered circuits.

Table 7.3 summarizes the comparison between the 4T CDMR memory and the other considered bitcell solutions [8, 10, 20]. The table clearly shows the benefits of the proposed solution, achieving much lower power and area consumption compared to other memory options and including inherent error detection capability and a low overhead error correction capability.

7.4 Conclusions

This chapter proposed a 4T GC-eDRAM with internal feedback for use in scaled CMOS nodes characterized by high leakage currents. The bitcell design protects the weak data level ("0") by a conditional, cell-internal feedback path, while the feedback does not protect the strong data level ("1"). The proposed cell is shown to enable low retention power of 3.86 pW/bit with a worst-case retention time of 8.29 ms at 27 °C, and 53.78 pW/bit with retention time of 2.76 ms at 85 °C. The bitcell area is 0.708 μm^2, which is 40% smaller than a 6T SRAM in the same technology, making it an appealing high-density, low-leakage alternative.

To provide a solution for high radiation environments, we presented a soft-error tolerant embedded memory based on a 4T GC-eDRAM, with CDMR integrated into each bitcell. The implementation of the proposed array with a parity column for error correction enables subthreshold operation of down to 400 mV, with a cell area that is 53% lower than a conventional 6T SRAM and much smaller than alternative rad-hard bitcells in a low-power 65 nm CMOS technology, while providing highly reduced power consumption.

References

1. Atias L, Teman A, Fish A (2013) A 13T radiation hardened SRAM bitcell for low-voltage operation. In: 2013 IEEE SOI-3D-Subthreshold microelectronics technology unified conference (S3S), 7–10 Oct 2013, pp 1–2
2. Atias L, Teman A, Fish A (2014) Single event upset mitigation in low power SRAM design. In: 2014 IEEE 28th convention of electrical & electronics engineers in Israel (IEEEI), 3–5 Dec 2014. IEEE, New York, pp 1–5
3. Bagatin M, Gerardin S, Paccagnella A, Faccio F (2010) Impact of NBTI aging on the single-event upset of SRAM cells. IEEE Trans Nucl Sci 57(6):3245–3250
4. Bajura M et al (2007) Models and algorithmic limits for an ECC-based approach to hardening sub-100-nm SRAMs. IEEE Trans Nucl Sci 54(4):935–945
5. Baumann R (2005) Soft errors in advanced computer systems. IEEE Des Test Comput 22(3):258–266
6. Baumann RC (2005) Radiation-induced soft errors in advanced semiconductor technologies. IEEE Trans Device Mater Reliab 5(3):305–316
7. Calhoun B, Chandrakasan A (2006) Static noise margin variation for sub-threshold SRAM in 65-nm CMOS. IEEE J Solid-State Circuits 1(7):1673–1679
8. Calin T et al (1996) Upset hardened memory design for submicron CMOS technology. IEEE Trans Nucl Sci 43(6):2874–2878
9. Cannon EH, KleinOsowski A, Kanj R, Reinhardt DD, Joshi RV (2008) The impact of aging effects and manufacturing variation on SRAM soft-error rate. IEEE Trans Device Mater Reliab 8(1):145–152
10. Chertkow O, Pescovsky A, Atias L, Fish A (2015) A novel low power bitcell design featuring inherent seu prevention and self correction capabilities. J Low Power Electron Appl 5(2): 130–150
11. Chun KC, Jain P, Lee JH, Kim C (2009) A sub-0.9V logic-compatible embedded DRAM with boosted 3T gain cell, regulated bit-line write scheme and PVT-tracking read reference bias. In: Proceedings of IEEE symposium on VLSI circuits (VLSIC), pp 134–135
12. Chun KC, Jain P, Lee JH, Kim C (2011) A 3T gain cell embedded DRAM utilizing preferential boosting for high density and low power on-die caches. IEEE J. Solid-State Circuits 46(6):1495–1505
13. Chun KC, Zhang W, Jain P, Kim C (2012) A 2T1C embedded DRAM macro with no boosted supplies featuring a 7T SRAM based repair and a cell storage monitor. IEEE J Solid-State Circuits 47(10):2517–2526
14. Clark LT, Patterson DW, Hindman ND, Holbert KE, Maurya S, Guertin SM (2011) A dual mode redundant approach for microprocessor soft error hardness. IEEE Trans Nucl Sci 58(6):3018–3025
15. Dodd PE, Massengill LW (2003) Basic mechanisms and modeling of single-event upset in digital microelectronics. IEEE Trans Nucl Sci 50(3):583–602
16. Garg R, Jayakumar N, Khatri SP, Choi GS (2009) Circuit-level design approaches for radiation-hard digital electronics. IEEE Trans Very Large Scale Integr VLSI Syst 17(6):781–792
17. Giterman R, Teman A, Meinerzhagen P, Atias L, Burg A, Fish A (2016) Single-supply 3T gain-cell for low-voltage low-power applications. IEEE Trans Very Large Scale Integr VLSI Syst 24(1):358–362
18. Ibe E, Taniguchi H, Yahagi Y, Shimbo Ki, Toba T (2010) Impact of scaling on neutron-induced soft error in SRAMs from a 250 nm to a 22 nm design rule. IEEE Trans. Electron Devices 57(7):1527–1538
19. International technology roadmap for semiconductors (2011). http://www.itrs.net/Links/2011ITRS/Home2011.htm
20. Jahinuzzaman SM, Rennie DJ, Sachdev M (2009) A soft error tolerant 10T SRAM bit-cell with differential read capability. IEEE Trans Nucl Sci 56(6):3768–3773

21. Lee Y, Chen MT, Park J, Sylvester D, Blaauw D (2010) A 5.42 nW/kB retention power logic-compatible embedded DRAM with 2T dual-VT gain cell for low power sensing applications. In: Proceedings of IEEE Asian solid state circuits conference (A-SSCC), pp 1–4
22. Lutkemeier S, Jungeblut T, Berge HKO, Aunet S, Porrmann M, Ruckert U (2013) A 65 nm 32 b subthreshold processor with 9T multi-Vt SRAM and adaptive supply voltage control. IEEE J. Solid-State Circuits 48(1):8–19
23. Maiz J et al (2003) Characterization of multi-bit soft error events in advanced SRAMs. In: IEEE international electron devices meeting, 2003. IEDM '03 Technical Digest, 8–10 Dec. 2003, pp 21–4
24. Massengill LW, Bhuva BL, Holman WT, Alles ML, Loveless TD (2012) Technology scaling and soft error reliability. In: Proceedings of IEEE international reliability physics symposium (IRPS). IEEE, New York, pp 3C–1
25. Mavis DG, Eaton PH (2002) Soft error rate mitigation techniques for modern microcircuits. In: Proceedings of IEEE reliability physics symposium, pp 216–225
26. Meinerzhagen P, Teman A, Burg A, Fish A (2013) Impact of body biasing on the retention time of gain-cell memories. IET J Eng 1:1–4
27. Meinerzhagen P, Teman A, Giterman R, Burg A, Fish A (2013) Exploration of sub-VT and near-VT 2T gain-cell memories for ultra-low power applications under technology scaling. J Low Power Electron Appl 3(2):54–72. doi:10.3390/jlpea3020054
28. Mohr KC, Clark LT (2006) Delay and area efficient first-level cache soft error detection and correction. In: International conference on computer design, 1–4 Oct 2006. IEEE, New York, pp 88–92
29. Osada K, Yamaguchi K, Saitoh Y, Kawahara T (2004) SRAM immunity to cosmic-ray-induced multierrors based on analysis of an induced parasitic bipolar effect. IEEE J Solid-State Circuits 39(5):827–833
30. Shayan M et al (2012) SEU tolerant robust memory cell design. In: IEEE 18th international on-line testing symposium (IOLTS), 27–29 June 2012, pp 13–18
31. Somasekhar D, Ye Y, Aseron P, Lu SL, Khellah M, Howard J, Ruhl G, Karnik T, Borkar S, De V, Keshavarzi A (2009) 2 GHz 2 Mb 2T gain cell memory macro with 128 GBytes/sec bandwidth in a 65 nm logic process technology. IEEE J Solid-State Circuits 44(1):174–185
32. Sterpone L, Violante M (2005) Analysis of the robustness of the TMR architecture in SRAM-based FPGAs. IEEE Trans Nucl Sci 52(5):1545–1549
33. Teifel J (2008) Self-voting dual-modular-redundancy circuits for single-event-transient mitigation. IEEE Trans Nucl Sci 55(2):3435–3439
34. Teman A, Meinerzhagen P, Burg A, Fish A (2012) Review and classification of gain cell eDRAM implementations. In: Proceedings of IEEE convention of electrical and electronics engineers in Israel (IEEEI), pp 1–5
35. Teman A, Meinerzhagen P, Giterman R, Fish A, Burg A (2014) Replica technique for adaptive refresh timing of gain-cell embedded DRAM. IEEE Trans Circuits Syst II 61(4):259–263
36. Vashishtha V, Clark LT, Chellappa S, Gogulamudi AR, Gujja A, Farnsworth C (2015) A soft-error hardened process portable embedded microprocessor. In: IEEE Custom integrated circuits conference (CICC), 28–30 Sept 2015. IEEE, New York, pp 1–4
37. Wang HH, Chen YL, Yang CC, Lin CK, Jeng MC (2013) Compact modeling for application-specific high-sigma worst case. In: 2013 International conference on simulation of semiconductor processes and devices (SISPAD), pp 61–64. doi:10.1109/SISPAD.2013.6650574
38. Zhang K, Bhattacharya U, Chen Z, Hamzaoglu F, Murray D, Vallepalli N, Wang Y, Zheng B, Bohr M (2005) SRAM design on 65-nm CMOS technology with dynamic sleep transistor for leakage reduction. IEEE J Solid-State Circuits 40(4):895–901

Chapter 8
Conclusions

8.1 Summary

This book has proposed gain-cell embedded DRAM (GC-eDRAM) as a promising alternative to SRAM for the implementation of embedded memories in low-power VLSI SoCs. The presented GC-eDRAM circuits were targeted at a broad range of low-power VLSI SoCs, from ultra-low power systems operated at subthreshold (sub-V_T) voltages to power-aware high-performance systems operated at near-threshold (near-V_T) or nominal supply voltages. It was shown that the key to achieve energy efficiency in GC-eDRAM is a proper understanding and control of the factors that determine the data retention time and its statistical distribution.

Many novel circuit techniques to enhance the data retention time of GC-eDRAM were presented and thoroughly evaluated by circuit simulations and test chips, including a full transmission-gate (TG) write port, reverse body biasing (RBB), a replica technique, and partial internal feedback. Following an introduction to embedded memories and GC-eDRAMs, a detailed statistical retention time model was developed. Thereafter, conventional gain-cell (GC) topologies were analyzed and evaluated at (near-V_T) and then (sub-V_T) voltages in mature CMOS technologies, followed by novel GC topologies and assist techniques that were specifically invented and optimized for (near-V_T) and (sub-V_T) operation and implementation in deeply scaled CMOS nodes.

GC-eDRAM combines most of the advantages of SRAM and conventional 1T-1C eDRAM and avoids most of their respective drawbacks, making it an attractive option for the implementation of embedded memories. In fact, GCs are much smaller than SRAM bitcells (by as much as 50%), they exhibit a much lower bitcell leakage current than SRAM bitcells, they are fully compatible with standard digital CMOS technologies (like SRAM, and unlike 1T-1C eDRAM requiring extra process steps and additional costs to build high-density 3D capacitors), they allow for nondestructive read access and can avoid power-hungry restore (write-back) operations (as opposed to 1T-1C eDRAM), and they have a separate

P. Meinerzhagen et al., *Gain-Cell Embedded DRAMs for Low-Power VLSI Systems-on-Chip*, DOI 10.1007/978-3-319-60402-2_8

read and write port (unlike conventional 6T SRAM and 1T-1C eDRAM), which allows to simultaneously and independently optimize the bitcell for high read and write robustness and allows for low-overhead two-port memory macrocell implementations with high access bandwidth. The main drawback of GC-eDRAM compared to 1T-1C eDRAM is the lower in-cell storage capacitance, which is built using exclusively MOSCAPs, junction capacitances, and interconnect capacitances available in a digital CMOS process, as compared to the dedicated stacked or trench DRAM capacitors. This typically results in lower data retention times and more frequent, power-consuming refresh operations.

While almost all previous works on GC-eDRAM targeted large cache memories for high-end microprocessors, this book extended the application range of GC-eDRAM to low-power VLSI SoCs (such as biomedical implants, sensor networks, and power-aware wireless communications systems). In particular, this book pioneered the field of low-voltage operation for GC-eDRAMs, exploiting near-threshold (near-V_T) and even subthreshold (sub-V_T) circuit operation for low leakage power and low access energy, as well as voltage-compatibility with and integration into ultra-low voltage (ULV)/ultra-low power (ULP) VLSI systems.

An analytical model for the distribution of the retention time was derived based on statistics of primary circuit parameters, such as the threshold voltage and the in-cell gate, junction, and interconnect capacitances. While all primary circuit parameters follow a Gaussian distribution, it was shown that the per-cell retention time of GC-eDRAM follows a log-normal distribution. The presented simple analytical retention time distribution model is highly precise when compared to Monte Carlo circuit simulation and 0.18 μm CMOS silicon measurement results. A PB-DOE sensitivity analysis unveiled that the threshold voltage of the write transistor is the sole most dominant parameter whose uncertainty leads to large per-cell retention time spread.

The following sections summarize the techniques which were presented in this book to improve the retention time of near-V_T and sub-V_T GC-eDRAMs, as well as aggressively scaled GC-eDRAMs.

8.1.1 Near-V_T GC-eDRAM Techniques

In a 2-PMOS GC-eDRAM implemented in a mature 0.18 μm node, voltage scaling from the nominal voltage (1.8 V) to a near-V_T voltage (0.7 V) enhances the data retention time by 4× provided that write access is unlikely and that the write bit-line (WBL) can be controlled to ground during standby and read. The retention time can be further improved by 3.3× if the WBL is set to a voltage between the supply rails, which, however, comes at the cost of voltage generation circuits and is particularly interesting only for large GC-eDRAM arrays. Even with this total 13.2× improvement in retention time, the data retention power is still dominated by the active refresh power, while leakage power in the GC-eDRAM array plays only a minor role. Therefore, several techniques to further improve the retention time and reduce the active refresh power (thus significantly reducing the data retention power) of near-V_T GC-eDRAM have been proposed in this book.

First, a 3-transistor (3T) gain-cell exhibits a full transmission-gate instead of a single write transistor, and a conventional merged storage and read transistor. This gain-cell topology ensures fast write access at low voltages and avoids the use of any overdrive or underdrive voltage, thereby facilitating its integration into a digital SoC and avoiding the need for a costly voltage regulator.

Second, silicon measurements of a 2 kb GC-eDRAM macrocell implemented in a 0.18 μm CMOS process show that the retention time can be improved by 2.3× (from 23 to 53 ms) by applying a reverse body bias (RBB) of only 100 mV. This is the first demonstration of successfully applying reverse body biasing to GC-eDRAM arrays, which has only been used in conventional 1T-1C eDRAM thus far. Moreover, silicon measurements show that 100 mV forward body biasing (FBB), which can be selectively applied for fast memory access, leads to a 2.9× retention time penalty. Sweeping the body voltage over a range of 375 mV spans a retention time range of almost 2 orders-of-magnitude, providing an interesting trade-off between access time and retention time.

Third, a replica bitcell technique, also implemented on a 2 kb all-PMOS 2T GC-eDRAM array in 0.18 μm CMOS, successfully tracks the retention time of the GC-eDRAM array across process-voltage-temperature (PVT) variations and varying write-access disturb frequencies. Silicon measurements show that the implemented replica technique allows to trigger refresh cycles up to 5× less frequently compared to conventional worst-case design, which significantly reduces the refresh power.

8.1.2 Sub-V_T and Deeply Scaled GC-eDRAM Techniques

The possibility of operating GC-eDRAM at subthreshold (sub-V_T) voltages, for use in ultra-low power systems, and of implementing GC-eDRAM in deeply scaled CMOS nodes, for use in future high-performance VLSI systems, has been investigated in this book. In order to enable sub-V_T operation in mature, above-100 nm CMOS nodes, the main design goals of the bitcell are long retention time and high data integrity. In the considered 0.18 μm CMOS node, a low-leakage I/O PMOS write transistor and an extended storage node capacitance ensure a retention time of at least 40 ms. Since at ultra-low voltages the data integrity is severely threatened by charge injection and clock feedthrough (capacitive coupling from read and write word-lines), a core NMOS transistor is used as read transistor to balance the storage node (SN) voltage disturbs (positive for write, negative for read); in addition, the core NMOS device is the strongest among all possible device options, ensuring a fast read operation and high array availability (i.e., fast read compared to retention time). Monte Carlo simulations of an entire 2 kb memory array, based on this mixed sub-V_T gain-cell design, operated at 1 MHz with a 400 mV sub-V_T

supply voltage, confirm robust write and read operations under global and local parametric variations, as well as a minimum retention time of 40 ms leading to 99.7% availability for read and write.

In deeply scaled CMOS technologies, such as the considered 40 nm CMOS node, subthreshold conduction is still dominant at ultra-low supply voltages. Gate tunneling and GIDL currents are still small, but of increasing importance, while reverse-biased pn-junction leakage and edge-direct tunneling currents are negligible. In the 40 nm node, the write transistor is best implemented with an HVT core PMOS device, which provides the lowest aggregated leakage current from the storage node (SN), even compared to the I/O PMOS device. Among various NMOS read transistor options, a standard-V_T core device maximizes the sense current ratio between a weak "1" and a weak "0" for near-V_T supply voltages (600–800 mV), where 97% array availability is achieved. Both the access times and the retention time are roughly three orders-of-magnitude shorter than in the 0.18 μm CMOS node, due to the increased leakage currents and smaller storage node capacitance. Briefly, we showed the feasibility of sub-V_T GC-eDRAM operation for mature process technologies and near-V_T operation for a deeply scaled 40 nm process, and provided best-practice bitcell designs for achieving minimum V_{DD} at these two very different nodes.

In order to realize GC-eDRAM in deeply scaled CMOS nodes characterized by high leakage currents, this book presented a novel 4-transistor (4T) gain-cell which contains an internal feedback transistor to strengthen the weaker data level, while hardly affecting the stronger data level. This leads to a much more symmetric decay of data "0" and "1" and a significantly extended retention time compared to a baseline 2T gain-cell.

Technology and voltage scaling also leads to a rise in the susceptibility of embedded memories to radiation-induced soft-errors, often limiting their fabrication to older, more robust process technologies. While most approaches to improve the redundancy of embedded memories are based on SRAM, this book presented a redundant 4T GC-eDRAM bitcell, combining both circuit and architectural techniques to provide a soft error tolerant dynamic memory, designed for ULP applications. The redundant 4T GC-eDRAM array showed full compatibility with subthreshold operating voltages down to 400 mV in a 65 nm CMOS technology, with smaller cell size and lower power consumption than SRAM-based bitcell solutions.

8.2 Outlook

Compared to conventional 6T SRAM and 1T-1C eDRAM, gain-cell based eDRAM (GC-eDRAM) has a crucial advantage which can make it appealing for the implementation of embedded memories in advanced CMOS nodes or for operation at scaled voltages. In fact, as explained in detail in Chap. 2, gain-cells have a separate read and write port, which allows the simultaneous and independent optimization of a gain-cell for both robust read and write operations. Unfortunately, beside the possibility of achieving both robust read and write, the large spread of per-cell retention time and the small in-cell storage capacitor, coupled with conventional refresh time guardbanding, leads to power-hungry refresh cycles. Therefore, especially for large cache memories, where an extremely unlikely worst-case cell dictates the refresh period, the adoption of GC-eDRAM is not an attractive option for the major semiconductor companies, which, in turn, focus most of their research on innovative ways of obtaining large, dedicated DRAM capacitors in below 28 nm CMOS nodes. For example, recent patents of Intel propose to use the readily available fin structure, used to build FinFETs (tri-gate transistors) to build large and high-density capacitors.

We believe that combining the advantages of the gain-cell read and write ports with large, emerging, dedicated DRAM capacitors would lead to a winning new type of memory bitcell for future VLSI applications. However, if the use of dedicated DRAM capacitors is not economic, GC-eDRAM is still an interesting option for many VLSI SoCs requiring medium-size memory arrays and rather short data retention times. There are certainly many applications, such as the LDPC decoders mentioned in Chap. 2, which can benefit from GC-eDRAM, either operated with periodic refresh cycles, or in a refresh-free way due to frequent write updates. Besides such high-performance VLSI DSP systems, GC-eDRAMs are also an interesting memory option for the niche of future ultra-low power (ULP) VLSI systems operated at ultra-low voltages (ULV) and implemented in mature, low-leakage, low-cost CMOS processes (such as 0.18 μm CMOS). In fact, as seen in Chap. 6, it is possible to operate GC-eDRAM at sub-V_T voltages in mature CMOS nodes, and such sub-V_T 2T-bitcell GC-eDRAM is an extremely high-density alternative to the currently used 8-14T-bitcell sub-V_T SRAM macrocells.

The analytical GC-eDRAM retention time distribution model introduced in this book enables to model and exploit the trade-off between read failure probability and refresh power in future error-resilient low-power VLSI systems. In such future VLSI systems, the refresh rate could even be set dynamically in order to selectively change between an accurate, power-hungry computing mode and a less accurate, low-power computing mode. Research at the system level and, in particular, from a fault tolerance perspective could take significant advantage of the large per-cell retention time spread coupled with a dynamically set refresh rate.

Glossary

1T-1C	1-transistor-1-capacitor
2T	2-transistor
3T	3-transistor
4T	4-transistor
6T	6-transistor
8T	8-transistor
ASIC	Application-specific integrated circuit
BIST	Built-in self-test
BL(s)	Bit-line(s)
CDMR	Complementary dual-modular redundancy
CF	Clock feed-through
CI	Charge injection
CMOS	Complementary metal-oxide-semiconductor
C_{SN}	Storage node capacitor
DMR	Dual modular redundancy
DRAM	Dynamic random-access memory
DRT	Data retention time
DVFS	Dynamic voltage and frequency scaling
ECC	Error correction codes
eDRAM	Embedded DRAM
FPGA	Field-programmable gate array
GC	Gain cell
GC-eDRAM	GC based eDRAM
IFB	Internal feedback
LDPC	Low-density parity-check
LVT	Low threshold-voltage
MC	Monte Carlo
MOSFET	Metal-oxide-semiconductor field-effect transistor
MR, NR	Read transistor

P. Meinerzhagen et al., *Gain-Cell Embedded DRAMs for Low-Power VLSI Systems-on-Chip*, DOI 10.1007/978-3-319-60402-2

MW	Write transistor
Near-V_T	Near-threshold
NTC	Near-threshold computing
NTV	Near-threshold voltage
NW, PW	Write transistors
PCB	Printed circuit board
PVT	Process, voltage, temperature
RBB	Reverse body biasing
RBL	Read bit-line
RCLK	Read clock
RWL	Read word-line
SA	Sense amplifier
SEU	Single event upset
SN	Storage node
SNB	Complementary storage node
SoC	System-on-chip
SRAM	Static random-access memory
Sub-V_T	Subthreshold
TG	Transmission-gate
TMR	Triple modular redundancy
ULP	Ultra-low power
ULV	Ultra-low voltage
V_{DD}	Supply voltage
VLSI	Very large scale integration
V_T	Threshold voltage
WBL	Write bit-line
WL	Word-line
WWL	Write word-line
WWLn, WWLp	Write word-lines

Index

P. Meinerzhagen et al., *Gain-Cell Embedded DRAMs for Low-Power VLSI Systems-on-Chip*, DOI 10.1007/978-3-319-60402-2

MIX
Papier aus verantwortungsvollen Quellen
Paper from responsible sources
FSC® C105338

If you have any concerns about our products,
you can contact us on
ProductSafety@springernature.com

In case Publisher is established outside the EU,
the EU authorized representative is:
Springer Nature Customer Service Center GmbH
Europaplatz 3, 69115 Heidelberg, Germany

Printed by Libri Plureos GmbH
in Hamburg, Germany